Molten Color

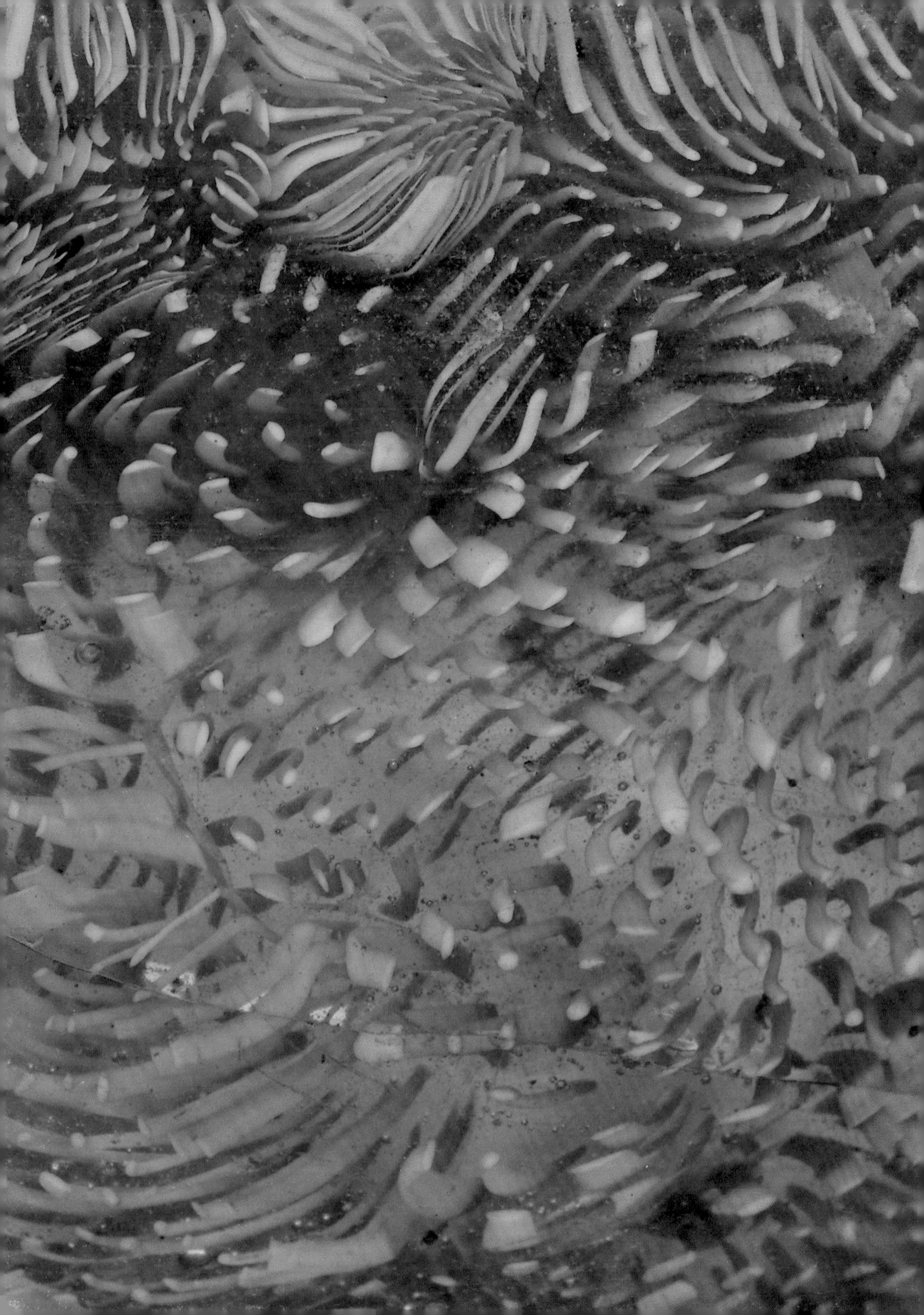

Molten Color

Glassmaking in Antiquity

Karol B. Wight

Published by the J. Paul Getty Museum

Getty Publications
1200 Getty Center Drive, Suite 500
Los Angeles, CA 90049-1682
www.gettypublications.org

Gregory M. Britton, *Publisher*

Ann Lucke, *Editor*
Elizabeth Chapin Kahn, *Production Coordinator*
Stuart Smith, *Designer*

Printed in China

Front cover: Ribbon flask. Roman, first century A.D. (see Figure 28)
Back cover: Hemispherical mosaic bowl. Late Hellenistic or Roman, first century B.C. (see Figure 23)

Library of Congress Cataloging-in-Publication Data

Wight, Karol 1958–
Molten color : glassmaking in antiquity / Karol B. Wight.
  p. cm.
 Includes bibliographical references and index.
 ISBN 978-1-60606-053-7 (pbk.)
1. Glass manufacture—History. 2. Glassware, Ancient. I. Title.
 TP850.W54 2011
 748.2—dc22
                          2010028704

CONTENTS

ACKNOWLEDGMENTS

I first became interested in ancient Roman glass when I started a graduate internship in the Department of Antiquities at the J. Paul Getty Museum in 1985. I was asked to prepare some acquisition proposals for glass vessels coming up at auction, a task that involved researching each piece in order to understand its function, manufacture, relevance, and relationship to the Museum's permanent collection. From that moment I was hooked, and within a few weeks I had determined that I would write my Ph.D. dissertation on one of those vessels, a piece that was successfully purchased at that auction and became part of the permanent collection (see Figure 60).

Along the way to becoming a specialist in ancient glass, I was encouraged and assisted by a number of colleagues, including Dr. Susan Downey, my graduate advisor and the cochair of my dissertation committee, and Dr. David Whitehouse, executive director of the Corning Museum of Glass and the other chair of my committee.

The glass world is a small one compared to other disciplines, one full of fascinating individuals who study, create, and collect works of glass art. My enthusiasm for this field of study that was generated in 1985 remains undiminished to this day, and I took great pride and joy in installing the *Molten Color* exhibition at the Getty Villa as one of its inaugural exhibitions when the Villa reopened in January 2006. This book stems directly from that exhibition, and I thank Greg Britton, publisher of the Getty Museum, for suggesting that I take on this writing project.

I would like to thank David Saunders and Travis Miles for their assistance in securing images of the Museum's objects and in answering a number of detailed questions about object dimensions and other matters. The stunning images of the works in the collection were taken by Ellen Rosenbery and Tahnee Cracchiola. Other images were provided by The Corning Museum of Glass and the Römisch-Germanisches Museum in Cologne. Jerry Podany, Erik Risser, and Marc Walton fielded arcane questions about analytical methods and production techniques.

Peter Evans, Lisa Bechtold (former senior staff assistant in the Department of Antiquities), and Minyoung Park kept this project and my life organized. In Getty Publications, thanks are owed to Ann Lucke, Catherine Chambers, Elizabeth Chapin Kahn, and Stuart Smith for producing this beautiful volume. My thanks go out also to my anonymous reviewer for posing a number of interesting questions that forced me to get back into researching glass topics, a task I always enjoy.

And finally, to my wonderful husband, Steven Wight, my best critic and editor par excellence, I dedicate this volume.

ABBREVIATIONS

ca.	circa
cm	centimeter
CMOG	Corning Museum of Glass, Corning, New York
D	depth
DIAM	diameter
GRI	Getty Research Institute, Los Angeles
H	height
in.	inch
inv. no.	inventory number
JPGM	J. Paul Getty Museum, Los Angeles
L	length
max.	maximum
RGM	Römisch-Germanisches Museum, Cologne
W	width

INTRODUCTION

A plate, a cup, a window, a mirror, a lens, a windshield, an eyedropper, a shower door, a test tube, a necklace, a chandelier, a pitcher. We are surrounded in our lives every day by objects made of glass; it is such a part of our lives that we rarely take a moment to consider what glass is or how it has been formed into the shapes it takes. Most of us have no idea that glassmaking began over three thousand years ago or that the techniques developed over two thousand years ago to shape it into a variety of pleasing and useful forms are the same techniques that are still employed by glass artists today. In my office sits a large bowl made of blue, clear, and white glass in a swirling, marbled pattern. The shape and function of such objects haven't changed over the millennia, and the colorful design of this bowl would be at home in an ancient Roman house of the first century A.D. *Plus ça change, plus c'est la même chose.*

Like stone and stonecutting, or clay and throwing pots on a potter's wheel, glassmaking is accepted in modern society as something that is and always has been. When visitors to the Getty Villa encounter a glass object from antiquity, they are usually amazed that something so fragile has survived for thousands of years. They marvel at the colors of ancient glass and at the fanciful shapes of some vessels (Figure 1). Perhaps one explanation for this habit of taking things for granted is that today we use glass in much the same way it was employed in antiquity. We utilize glass for drinking vessels, for windows and mirrors, for lenses in our eyeglasses, for jewelry, for chandeliers, and for many other objects.

Of course there have been developments in glassmaking over the past few centuries. One is **tempered glass** (see the Glossary of Glass Terms for more information on special terms), glass that when broken does not shatter into a thousand sharp shards that can cut or maim but rather stays together or simply crumbles. We use tempered glass for cooking and baking (how many Pyrex© containers are in your kitchen cupboards?), as well as for tables, car windows, and shower doors. Another example of a new application for glass is fiber-optic cable. Made from extremely thin rods of glass that carry light along their lengths and transmit

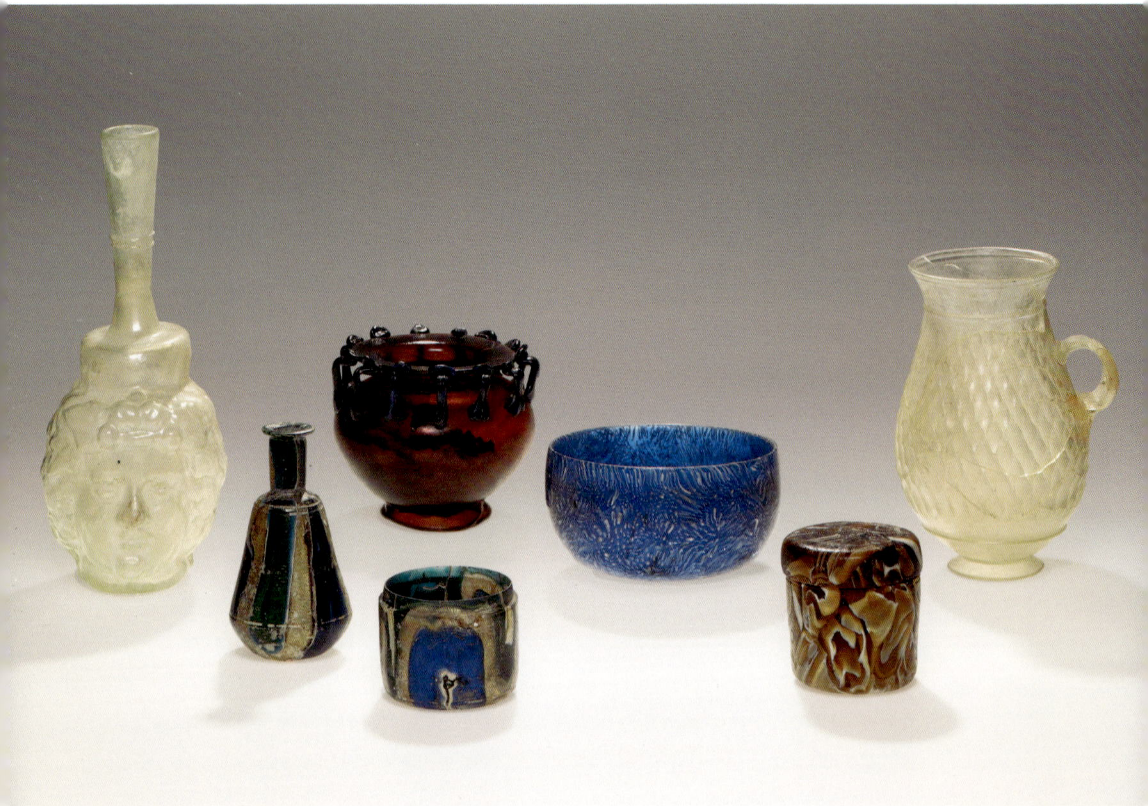

FIGURE 1

A group of ancient glass vessels made using a variety of techniques (*left to right*).
Mold-blown head flask. Roman, ca. second century A.D. (see Figure 56).
Gold band perfume flask. Late Hellenistic or Roman, 25 B.C.–A.D. 25. H 9.1 cm (3⁹⁄₁₆ in.).
JPGM 2003.230.
Vessel with thirteen handles. Roman, third–fourth century A.D. H 8.2 × DIAM 9.7 cm (3¼ × 3¹³⁄₁₆ in.).
JPGM 2003.398.
Gold band container (pyxis). Late Hellenistic or Roman, 25 B.C.– A.D. 25 (see Figure 82).
Mosaic bowl. Late Hellenistic or Roman, first century B.C. (see Figure 23).
Mosaic container (pyxis) with lid. Roman, first century A.D. (see Figure 27).
One-handled jug with facet-cut decoration. Roman, first–second century A.D. H 13.2 cm (5³⁄₁₆ in.).
JPGM 2003.346.

signals more effectively than metal wire, fiber-optic cables also can be bundled together as a source of artificial light. One of the reasons that the properties of glass continue to be explored and exploited is because today we have better control over the elements that allow glass to be manipulated. Although the ancients were able to employ sophisticated methods to make certain types of glass, such as **dichroic** or cameo (see chapter 4), today there is an even greater refinement of the individual components used to make glass, as well as the more precise regulation of the energy sources used to fuse these elements into glass. We are also able to manufacture glass objects more quickly through industrialized processes introduced in the late nineteenth and early twentieth centuries—all the jars and bottles that contain our jams, jellies, condiments, and beverages, for example, are formed on machines. In an assembly-line process, hot glass is introduced into a mold; compressed air is then used to create an internal cavity that forces the glass against the sides of the mold to take its design and shape, creating a space for the contents that will ultimately be placed inside. While glass workers in industrial settings were able to fashion a number of identical vessels quickly using shaped molds, they could not compete with the speed of such machines or with the volume of glass vessels that they produce.

In writing this book, I have the hope that it will transport the reader back in time four millennia, to about 2000 B.C., in order to follow the history of glass from its earliest stages of development in Mesopotamia and Egypt and on through the next two millennia to see how the process of glassmaking changed in antiquity as glassmakers became more proficient and developed new ways to work the material. What they accomplished thousands of years ago created the glassmaking legacy with which we live today.

CHAPTER 1

There is a story that once a ship belonging to some traders in natural soda [natron] put in here [at the mouth of the Belus River] and that they scattered along the shore to prepare a meal. Since, however, no stones suitable for supporting their cauldrons were forthcoming, they rested them on lumps of soda from their cargo. When these became heated and were completely mingled with the sand on the beach a strange translucent liquid flowed forth in streams; and this, it is said, was the origin of glass.

Pliny the Elder, Natural History 36.65.191

What Is Glass?

This Latin passage was written by Gaius Plinius Secundus, known more commonly as Pliny the Elder (A.D. 23–79). Pliny was a Roman navy and army commander, as well as an author and naturalist. His lifelong observations of the natural world led to the writing of an encyclopedic multivolume series called the *Naturalis Historia* (*Natural History*) that was issued between A.D. 77 and 79. In Book 36, quoted above, Pliny recounts a tale about the accidental discovery of glass by natron merchants. Pliny's tale is refuted by actual archaeological evidence from Mesopotamia, where vitreous glazes had been in use long before those natron merchants put to sea. It is generally accepted that the process of making glass was first discovered in the Bronze Age, toward the middle of the third millennium B.C. The exact location of the initial glassmaking industry is uncertain, but archaeological evidence suggests western Asia, where Sumerian archaeological sites have yielded lumps of raw glass or glass objects from contexts as early as the twenty-third and twenty-first centuries B.C. At that time, vitreous glazes were used to color beads, tiles, pottery, and other objects, and it is likely that the invention of glassmaking resulted from experiments with these colorful glazes (Figure 2).

So, putting folklore aside, what was known in antiquity (and continues to hold true today) is that all glass is a material made from three primary ingredients: silica, an alkali, and lime. Silica, which forms the largest percentage of the mixture, can come from sand, crushed quartz, or crushed flint (Figure 3). The silica used by today's glass artists is highly refined, but in antiquity glassmakers used sand found in riverbeds or along the seashore, and it often contained broken bits of shell from the crustaceans and mollusks living in the marine environment. These broken shells introduced lime into the **batch** (mix) of ingredients. Lime—calcined limestone—served to stabilize the batch. An alkali was needed to act as a **flux** to lower the melting point of the silica. Frequently it was a soluble salt leached from the ashes of burned plants, sometimes called potash or soda ash, which contain potassium or sodium carbonate. The mineral natron, composed of different sodium salts, was often used as an alkali in the Roman period. When these three basic components are heated and fused together to a temperature of about 1300°C–1500°C (2400°F–2700°F), they form glass. Since every recipe for making glass includes these ingredients, it is nearly impossible to categorize glass vessels by region or workshop based solely on an analysis of the chemical content of glass. "Natural" glass—that is, glass to which no specific colorant has been added—has enough trace minerals, such as iron in the sand used to make it, to give it a greenish, yellowish, or bluish tinge, and many utilitarian Roman glass vessels have this trace of color. But to deliberately color the glass, minerals were added to the batch. Referring again to Pliny the Elder, we find that "Glass, like copper, is smelted in a series of furnaces, and dull black lumps are formed.... After being reduced to lumps, the glass is again fused in the workshop and is tinted" (*Natural History* 36.65.193). Thus, in accordance with Pliny's observations, color was added to the glass only after the raw ingredients had been heated and fused. The freshly made glass was then reheated and the colorant added to the batch. See Table 1 for the colors that result from the use of certain additives.

The process of introducing these mineral additives to the hot glass was a tricky business; as with firing ceramics, the amount of oxygen introduced into

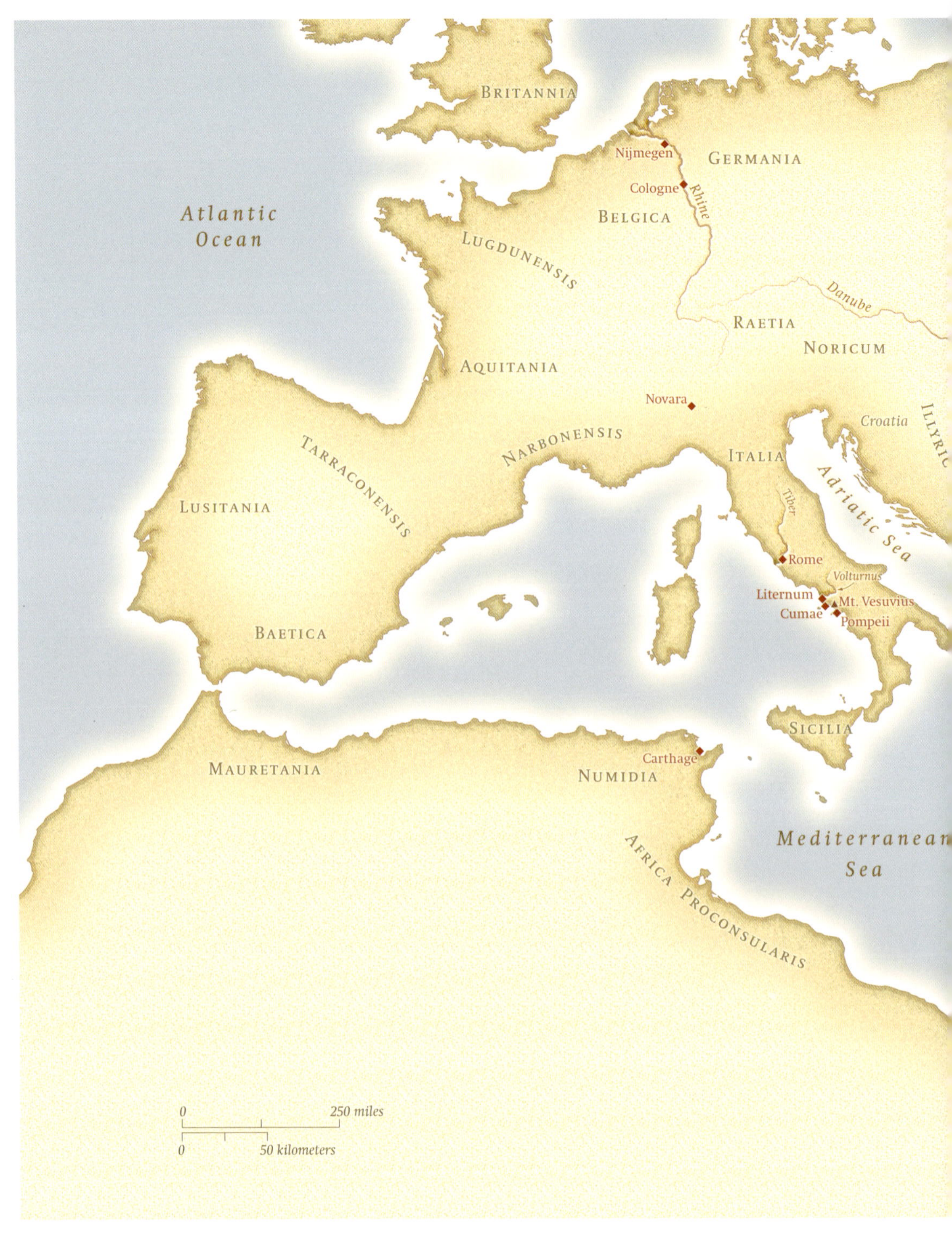

FIGURE 2
The Roman Empire, ca. A.D. 117.

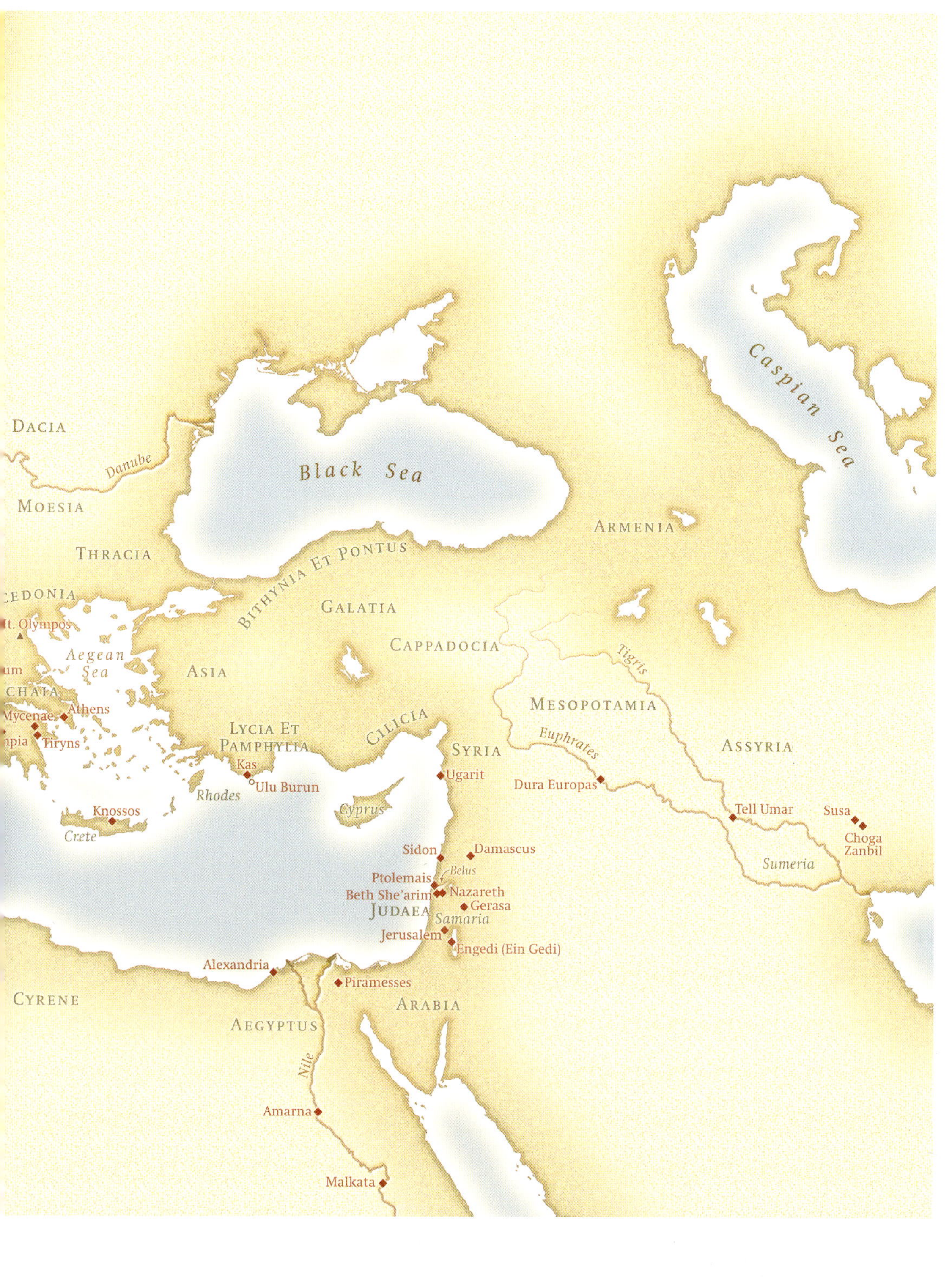

DACIA
MOESIA
THRACIA
CEDONIA
Mt. Olympos
Aegean Sea
um
CHAIA
Mycenae
Athens
pia
Tiryns
Knossos
Crete
CYRENE
Danube
Black Sea
BITHYNIA ET PONTUS
GALATIA
CAPPADOCIA
ASIA
LYCIA ET PAMPHYLIA
CILICIA
Kas
Ulu Burun
Rhodes
Cyprus
Sidon
Ptolemais
Belus
Beth She'arim
Nazareth
JUDAEA
Gerasa
Samaria
Jerusalem
Engedi (Ein Gedi)
Alexandria
Piramesses
ARABIA
AEGYPTUS
Nile
Amarna
Malkata
Ugarit
SYRIA
Damascus
Dura Europas
Euphrates
MESOPOTAMIA
Tigris
ARMENIA
Caspian Sea
ASSYRIA
Tell Umar
Susa
Choga Zanbil
Sumeria

the kiln played a major role in determining the final color. So, for example, allowing air to reach a batch of glass to which copper had been added would turn the final product from red to blue or green. One is left to wonder how many of the brilliantly colored glass vessels we see today were intentional—or accidental.

Beginning in the late Bronze Age, glass began to be manufactured in Mesopotamia and Egypt both in the form of usable objects, such as vessels or pendants, or shaped as ingots for transport as a raw commodity around the Mediterranean (see chapter 2). But in addition to glass made fresh from its primary ingredients, glass was also made by reheating recycled material. In his lampoon of Caecilius, Martial (Marcus Valerius Martialis, A.D. 38/41–102/104), the Roman poet and satirist, compares him to a peddler: "A wit, Caecilius, you fancy yourself. You are none, believe me. What then? A buffoon. You are just like the tramping hawker from beyond the Tiber who exchanges pale sulphur matches for broken glass" (*Epigram on Caecilius*, 1.41). Martial's predecessor, the poet Statius (Publius Papinius Statius, ca. A.D. 45–96), also equates vendors of sulphur matches with broken glass: "Here is the mob of the stage and vendors of common sulphur for broken glass" (*Silvae*, 1.6.74). One of the benefits of adding broken glass to the batch to make new vessels was that when it was added to glass either made from

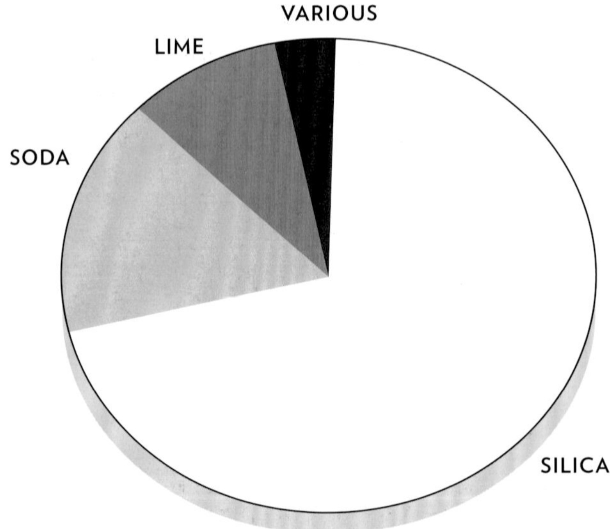

FIGURE 3
Components in the manufacture of glass.

fresh ingredients or reheated from ingots, it melted and fused at a lower temperature than glass made only from raw ingredients; thus a glass kiln needed to reach a temperature of only about 750°C (1380°F). This was an important economic factor for the proprietor of a glass workshop, who had to purchase the fuel to fire a glass kiln and who wanted to spend the least amount of money in order to do so. Adding recycled glass to a batch also encouraged cohesion.

Once heated and fused, the viscous liquid could be manipulated in a variety of ways using various tools, molds, and **blowpipes**. Unfortunately for us, few of the tools used in antiquity to fashion glass have survived, although they did leave their marks on the surface of the glass. Glass scholars have studied these marks and also turned to later descriptions and images of glassmaking to try to understand ancient methods. One of these resources is a series of illustrations in a remarkable encyclopedia written by Denis Diderot and Jean Le Rond d'Alembert, *Recueil des planches, sur les sciences et les arts liberaux et les arts méchaniques, avec leur explication*. Published in Paris in 1772, Diderot's *Encyclopédie*, as it is commonly known, provides images of the tools and techniques used by glassmakers in Paris in the eighteenth century. An illustration of some of the tools is very instructive (Figure 4). It shows the hollow and solid rods—the blowpipe and the **pontil**—which are necessarily long in order to protect the glassmaker from the heat of the kiln as he reaches in to remove hot glass and to distance himself from the end of the rod while a vessel is shaped, lessening the chances that he will be burned. Various wooden molds are shown that are used to help uniformly shape a bubble of glass, as are the pincers, **shears**, and **jacks** used to cut, push, and pull the glass into shape. A glassmaker used these tools while seated at a **chair** (Figure 5). There is no

TABLE 1. Chart of Mineral Additives and Resultant Color

Color	Additive
Colorless	Antimony or manganese
Purple	Manganese
Blue and green	Cobalt, copper
Opaque Red	Copper
White and yellow	Antimony, lead

(See Stuart J. Fleming, *Roman Glass: Reflections on Cultural Change* [Philadelphia: University of Pennsylvania Museum of Archaeology and Anthropology, 1999], appendix A, p. 137ff., for a more thorough chemical discussion of the additive elements.)

archaeological evidence to show that Roman glassmakers also used a chair, but such an aid would surely have helped. The benchlike structure has short arms at either end that are used to support the blowpipe or pontil while the vessel is being fashioned. When fashioning a vessel, a glassmaker must constantly rotate the rod in order to maintain the symmetry of the vessel's shape while the glass is in a viscous state. If the glassmaker fails to do this, the vessel will slump and be ruined before it is completed. The chair's arms also allow the glassmaker to rest the rod, rather than having to hold it continuously while using jacks, shears, or pincers to squeeze, push, or pull the glass into the desired shape. Modern glass artists who have attempted to replicate ancient techniques have strapped wooden strips to their thighs in order to imitate the function of the chair's arms, by providing on their body a sturdy spot to rest their rod. (This method can be seen on the website Roman Glassmakers, http://www.romanglassmakers.co.uk/furnace14.htm.)

When the glassmaker finished making his vessel, his product needed to be cooled in a controlled manner. Ancient glassmakers quickly discovered that the strain introduced into the hot glass during its manipulation could result in thermal shock that could shatter their work if it were cooled too quickly. The finished glass object first had to be reheated to a uniform temperature and then slowly cooled, a process called **annealing**. Glass kilns had a chamber separate from the one used to melt and fuse glass to hold the vessels being annealed. A fifteenth-century manuscript of *Sir John Mandeville's Travels* records the fantastic voyages of an English character named Sir John Mandeville as he journeyed through Europe and the Middle East, and even as far as China. Most interesting for our discussion here, the manuscript's illustrator added an image of glassmakers industriously at work (Figure 6). All aspects of the process are shown, from a worker shoveling sand to bring to the glass workshop, to an artisan inflating glass at the end of a blowpipe. The kiln is also clearly drawn. It has an opening at the lower right where fuel is added to keep the glass hot; two higher holes in the center (the **glory holes**) in which the molten glass sat in crucibles ready to be removed and fashioned into a vessel; and, at the left, the annealing chamber. The worker illustrated there is probably adding hot vessels. He would have needed to use a tool to place them in the annealing oven as the still-high temperature of a just-finished glass object would severely burn the hands of anyone trying to touch it. Once annealed, the glass was checked for any flaws and then sent to market, ready for use.

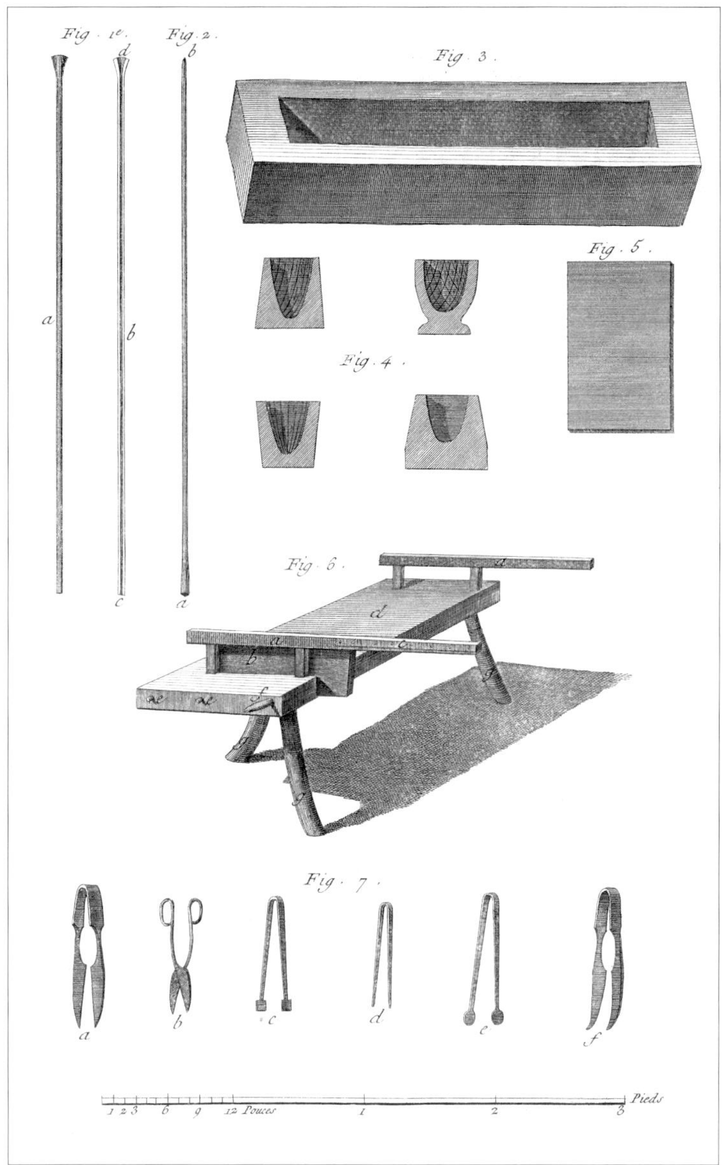

FIGURE 4

Verrerie en bois, différens outils pour le travail du verre (Different tools for the working of glass).
In this illustration, fig. 1 shows a blowpipe (a) and its transection (b), fig. 2 is a pontil, fig. 3 is a
wood box used to collect glass cuttings trimmed from vessels during manufacture, fig. 4 depicts
molds, fig. 5 is a marver, fig. 6 is a chair, and fig. 7 illustrates various sprung tools such as pincers,
shears, and jacks. From Denis Diderot and Jean Le Rond d'Alembert, *Recueil des planches, sur
les sciences et les arts liberaux et les arts méchaniques, avec leur explication* (Paris, 1772), vol. 10,
pl. 18. Los Angeles, Research Library, GRI no. 84-B31322. Photo: GRI.

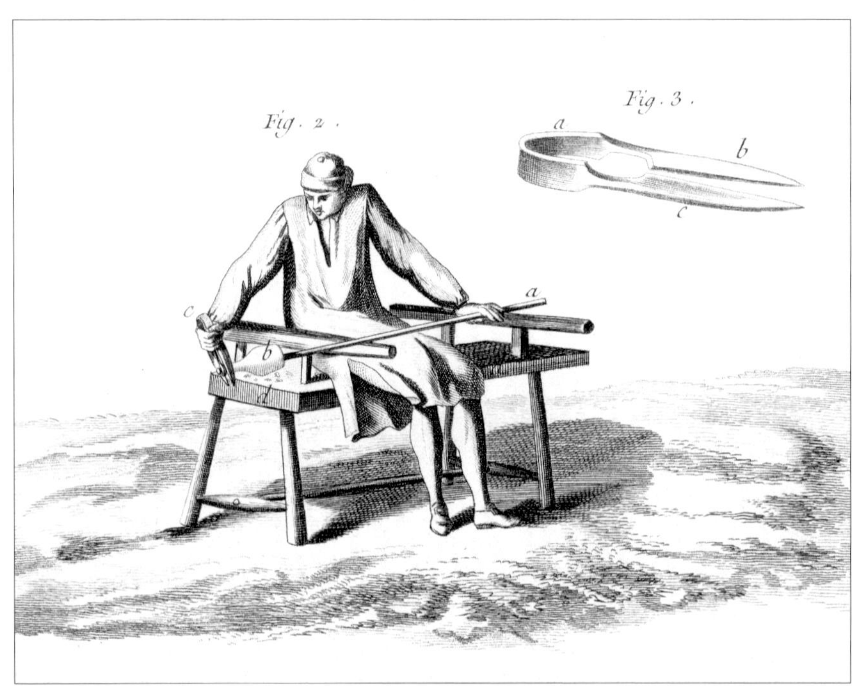

FIGURE 5

Chair. The craftsman is sitting on a chair (d) and using its arms to support the pontil (a) that he rolls back and forth while he forms the neck of the bottle (b) with jacks (c). From Denis Diderot and Jean Le Rond d'Alembert, *Recueil des planches, sur les sciences et les arts liberaux et les arts méchaniques, avec leur explication* (Paris, 1772), vol. 10, pl. 6, fig. 2. Los Angeles, Research Library, GRI no. 84-B31322. Photo: GRI.

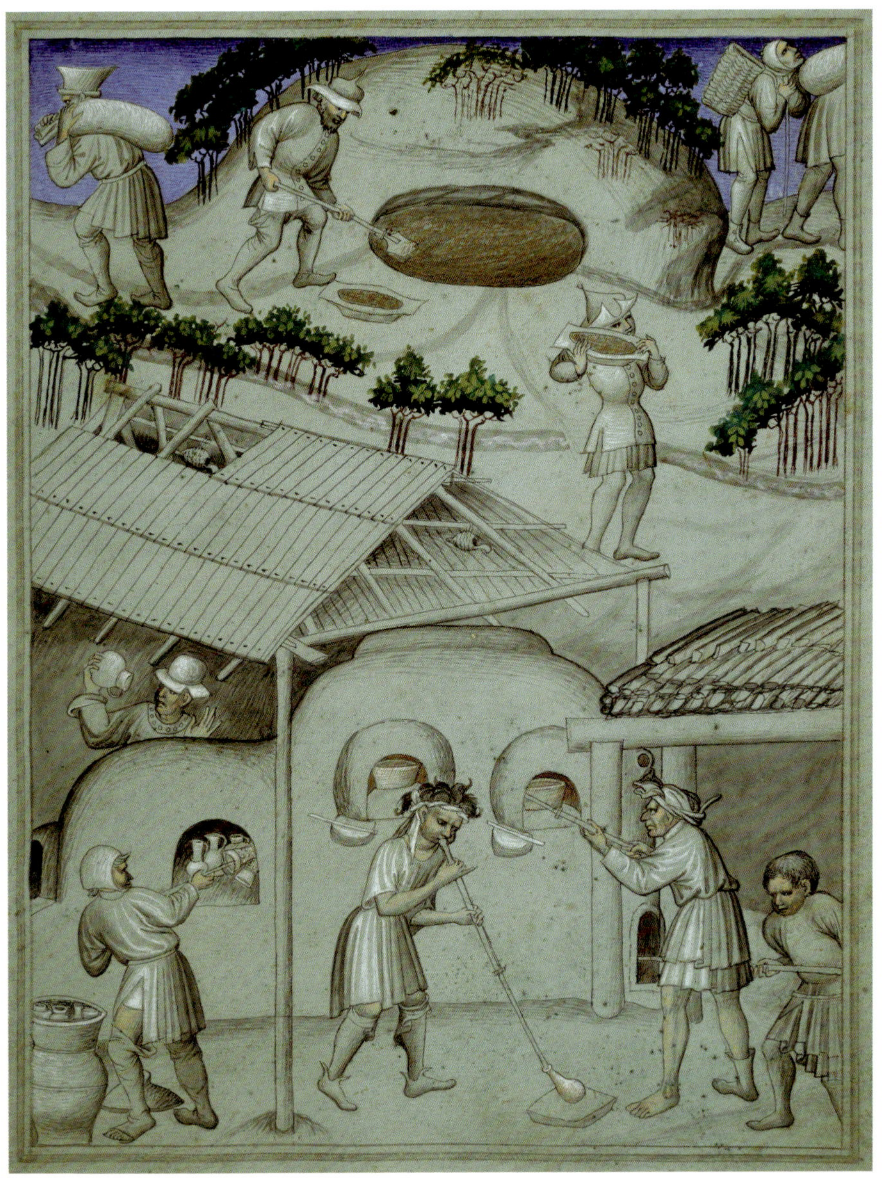

FIGURE 6

Glassmakers at work. At the top, a worker shovels sand. In the foreground, a glassblower inflates a glass bubble on a marver set on the ground; to the far right, a young man stokes the fire; to the left, another man uses a stick to add hot objects to the annealing chamber. On the other side of the furnace, a man sorts through cooled pieces. Beyond the glasshouse, wood ash is being burned and carted away; at the upper right, two men carry baskets filled with glassware to market. From *Sir John Mandeville's Travels* (detail of illustration), probably Bohemia, ca. 1420–50. London, The British Library, Ms. 24189, fol. 16v. Photo: The British Library.

CHAPTER 2

Our understanding of the early glass industry comes from two different, but related, disciplines: archaeology and philology. Archaeology is the discovery and interpretation of *physical* remains; philology, the interpretation of *literary* remains. These sources combined have shown that there were two early centers for the production of glass, Mesopotamia and Egypt.

The earliest archaeological evidence comes both in the form of excavated glass objects and in the sites where glass was manufactured. In antiquity, as today, raw materials were transported around the Mediterranean by sea and by land in order to be reshaped into a manufactured product. Glass was also a raw material and was conveyed in the form of ingots as early as the fourteenth century B.C. In 1984, underwater archaeologists from the Institute of Nautical Archaeology

The Birth of
Glass
Mesopotamia
and Egypt

began excavating a shipwreck near Kas off the coast of southern Turkey. Now known as the Uluburun shipwreck, it has been dated to the fourteenth century B.C. and contained a mixture of raw materials and finished merchandise. Part of its cargo consisted of about 175 disc-shaped glass ingots, the majority of which were cobalt blue or turquoise, but with one unique example in a lavender color. These ingots were being transported to another market where the glass could be reheated and formed into usable objects. Archaeologists believe that the ship may have sailed from a port along the eastern Mediterranean coast, perhaps from Ugarit. To date, the only late Bronze Age industrial sites associated with distributing glass to other parts of the Mediterranean have been excavated in Egypt at the sites of Malkata, Amarna, and Piramesses (near modern Qantir). In fact, molds used for making ingots similar to those found on the Uluburun shipwreck were found at Piramesses. But inscriptional evidence (discussed below) strongly suggests that factories for glassmaking also existed in Mesopotamia, although none have been found to date.

This incomplete archaeological record makes it difficult to identify the source of the raw glass used to manufacture the various objects that have been found in excavations. Scientists, therefore, have turned to analyzing the trace elements of finished glass objects to try to identify where their source glass was manufactured. Some of the most promising research is focusing on the trace elements associated with heavy minerals and clay (titanium, chromium, zirconium, and lanthanum). Identifying these trace elements has proven to be an effective means of distinguishing between Egyptian and Mesopotamian glass by determining the source of the silica used to make the glass, whether sand or quartz. Silica sources were an important factor in antiquity since their proximity was necessary for the success of glass manufacture, and Pliny's tale of the natron merchants who first made glass was in reference to the sands of the Belus River, which meets the Mediterranean near the ancient Greek colony of Ptolemais. He also writes of the "white sand which forms in the River Volturnus" in Italy, "found along six miles of the seashore between Cumae and Liternum" (*Natural History* 36.65.194). As scientists continue their research and work to refine their efforts, it may be possible to better understand the circulation of glass ingots from Egypt and Mesopotamia throughout the eastern Mediterranean and the Aegean. This analytical work on silica sources is not limited to Bronze Age glass, however; it is also being undertaken to analyze glass made in later periods in order to understand similar economic relationships during the Roman Empire.

Another analytical technique being undertaken today involves the identification of lead isotopes in glass, which are then compared to the known sources of lead in the ancient Mediterranean. Lead isotope analysis has been used on Egyptian, Roman, and Byzantine glass.

The first literary references for glassmaking come from Mesopotamia; one of the earliest records is a clay tablet thought to be from Tell Umar that dates to between 1400 and 1200 B.C. Now in the British Museum (Department of Western Asiatic Antiquities, inv. no. 120960), it gives in cuneiform script a Babylonian recipe for making red glass that includes instructions for adding lead, copper, and antimony. Scholars who have studied the tablet note that the language in which the recipe is written is slightly obscure, which would have made it more difficult for the nonspecialist to read—perhaps an early example of protecting trade secrets.

CAST GLASS

The earliest glass objects were manufactured around 2500 B.C. by first molding and then cold-working the glass (Figure 7). Scholars believe that the early glassmakers lacked the technology to heat glass sufficiently so that it could be manipulated easily. The temperatures they were able to achieve left the glass in a state insufficiently viscous to pour into a mold; they therefore heated crushed or larger chunks of glass directly in the mold and probably pushed it into place with tools. Once it had annealed, the molded glass could be polished or cut to achieve its final design. The open-faced molds used to make these early pieces were for small objects such as pendants and inlays. Almost all the earliest glass objects are opaque and brilliantly colored, and they were likely made as substitutes for rare precious stones. Several blues resembling lapis lazuli and turquoise are known, as is an opaque red like the gemstone carnelian. Some of the earliest cast pendants from Mesopotamia are images of the goddess Astarte that are dated to about 1400–1200 B.C (Figure 8). These pieces were deliberately pierced so that they could be strung onto a necklace and worn. Around the same time in Egypt, cast pieces were used to enhance sculpture, as inlays to decorate stone and wood reliefs and furniture, and as jewelry (Figures 9–10). Scholars have theorized that the Egyptian glassmaking industry emerged fully about 1450 B.C., after the conquests of the Egyptian pharaoh Tuthmosis III in Syria, when it seems likely that workers from that region came to Egypt.

Cast-glass jewelry pendants were also manufactured in Greece during the Mycenaean period around 1400 B.C. (Figure 11). Pendants were discovered in burials at the sites of Mycenae and Tiryns and elsewhere in the Greek world, and almost all are cobalt blue. While many share the same rectangular shape, a variety of patterns is carved into the molds, including rosettes or wavelike lines. A group of Mycenaean glass beads at the Getty Museum exhibits six different patterns, and recent material analysis of them by scientists working in England, the United States, and Belgium has indicated that the glass used for their manufacture came

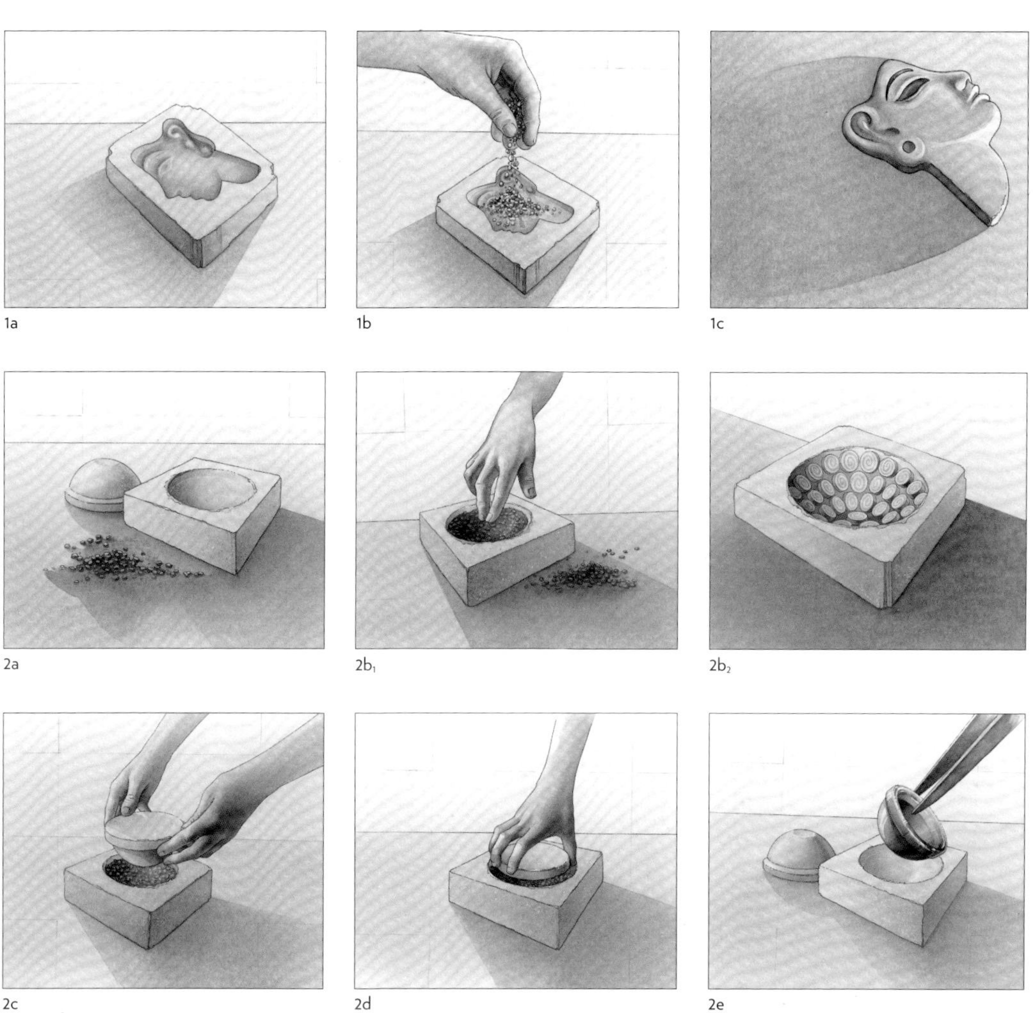

1a

1b

1c

2a

2b$_1$

2b$_2$

2c

2d

2e

FIGURE 7

Casting glass objects is a multistep process. To cast a pendant or inlay, crushed or chipped glass is placed into an open mold (1a–1b). When heated, the softened glass fills the mold. After annealing, the inlay or pendant is removed (1c). To form a vessel, the chipped glass, or segments of mosaic cane are placed within the mold (2a, 2b$_1$, 2b$_2$). A second mold part is inserted and will shape the inner wall of the vessel (2c). After heating, the inner mold is removed (2d), and the cast vessel is released from the mold for annealing (2e).

from Egypt. An open mold used to make this style of pendant was found at the site of Knossos on the island of Crete. Made of steatite, it is now in the collection of the Ashmolean Museum in Oxford, England (inv. no. 1910.522). Stone and terracotta, which could withstand the heat needed to melt the glass, were frequently used for molds.

Molds were also used to manufacture vessels, but unlike the open molds used for inlays, the vessel molds were designed to form both the exterior and the interior; the glass was melted between the walls formed by the molds in order to impart the proper shape to the vessel. Like the open molds used for inlays, the vessel molds were filled with either chipped glass or chunk glass, which was then heated and fused. These early monochrome pieces were cast as blanks, almost to the size desired, and then were manually cut, ground, drilled, and polished into their final shapes. Many of the most beautiful examples of cast and cut vessels come from the Achaemenid Empire of ancient Persia (Figure 12). Almost all are hemispherical bowls or shallow plates, and the shapes and patterns of the glass vessels are also found on precious silver tablewares. Characteristic of Achaemenid style, the bowls frequently have knobs on their exterior, perhaps to make them easier to hold. Nearly all the Achaemenid glass vessels are colorless, and it is likely that the glassmakers were trying to imitate rock crystal (Figure 13).

This technique of first casting and then cutting vessels to the desired shape before finally incising decorative patterns was used for centuries. The cutting and carving was achieved by means of various engraving tools and turning wheels combined with an abrasive. Such wheels were presumably the same ones used by gem engravers, carvers of hard stone vessels, and furniture makers—those familiar with cutting techniques—who could use their skill to decorate glass as well. The wheels were likely made of metal (modern wheels are copper), and when the turning wheel was combined with an abrasive material suspended in oil and dripped onto the wheel, the glass was ground away.

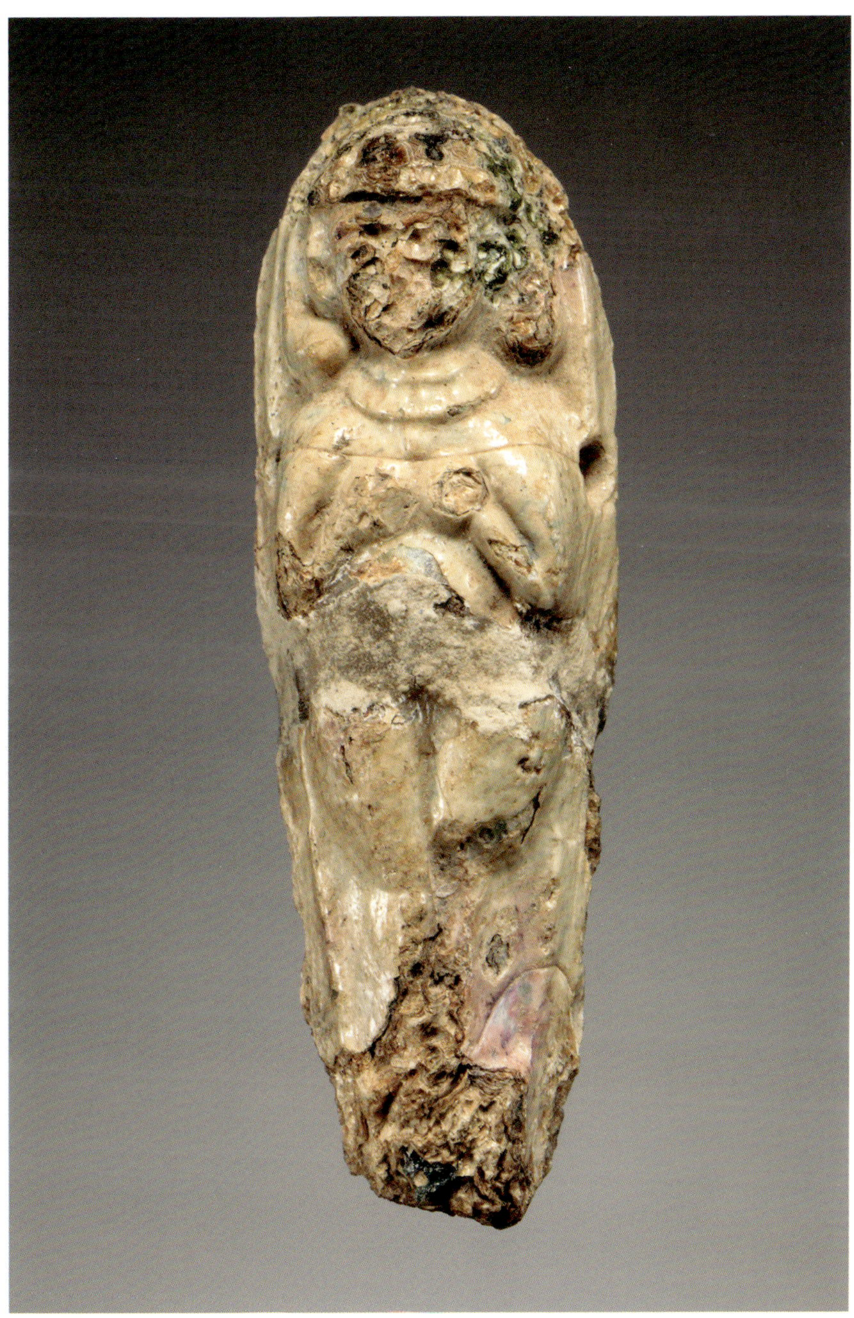

FIGURE 8
Cast pendant with the figure of Astarte. Mesopotamian, mid-second millennium B.C.
H 7.2 cm (2¹³/₁₆ in.). JPGM 2003.213

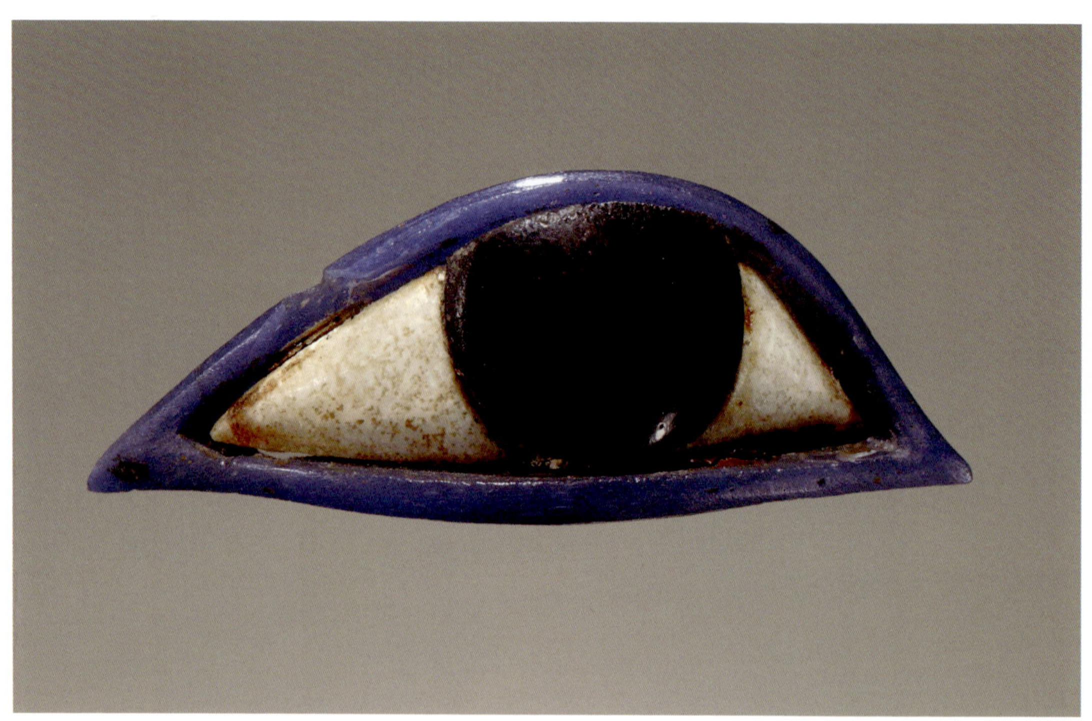

<hr>

FIGURE 9
Cast inlay in the form of an eye. Egyptian, New Kingdom, 1540–1070 B.C.
H 1.4 × W 4.6 × D 1 cm ($9/16$ × $1^{13}/16$ × $3/8$ in.). JPGM 2003.154.

FIGURE 10
Cast inlay in the form of a head. Egyptian, New Kingdom, 1540–1070 B.C.
H 3 cm ($1^{3}/16$ in.). JPGM 2003.149.

FIGURE 11 (pp. 22–23)
Group of cast pendant beads. Greek, Mycenaean, fourteenth century B.C.
L 2.5–4 cm (1–$1^{9}/16$ in.). JPGM 2003.214.

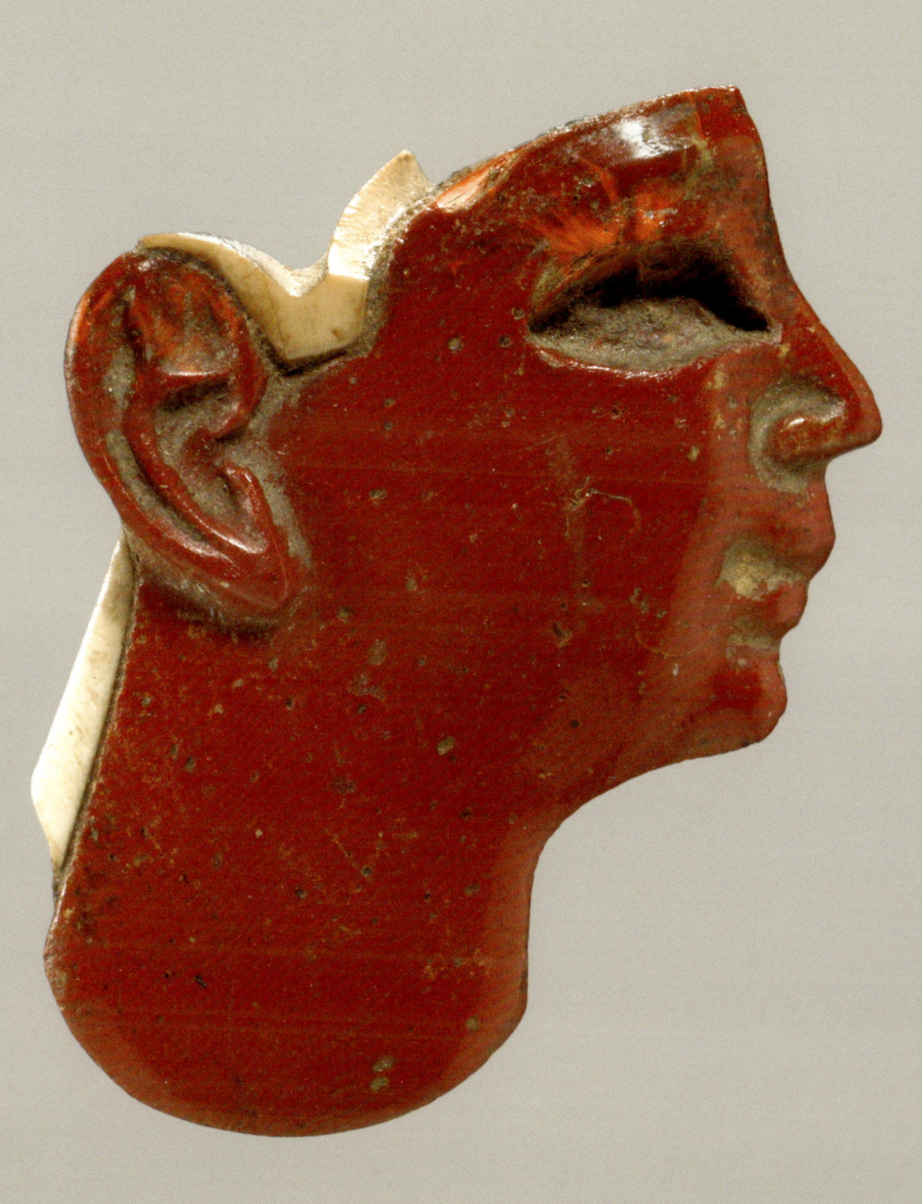

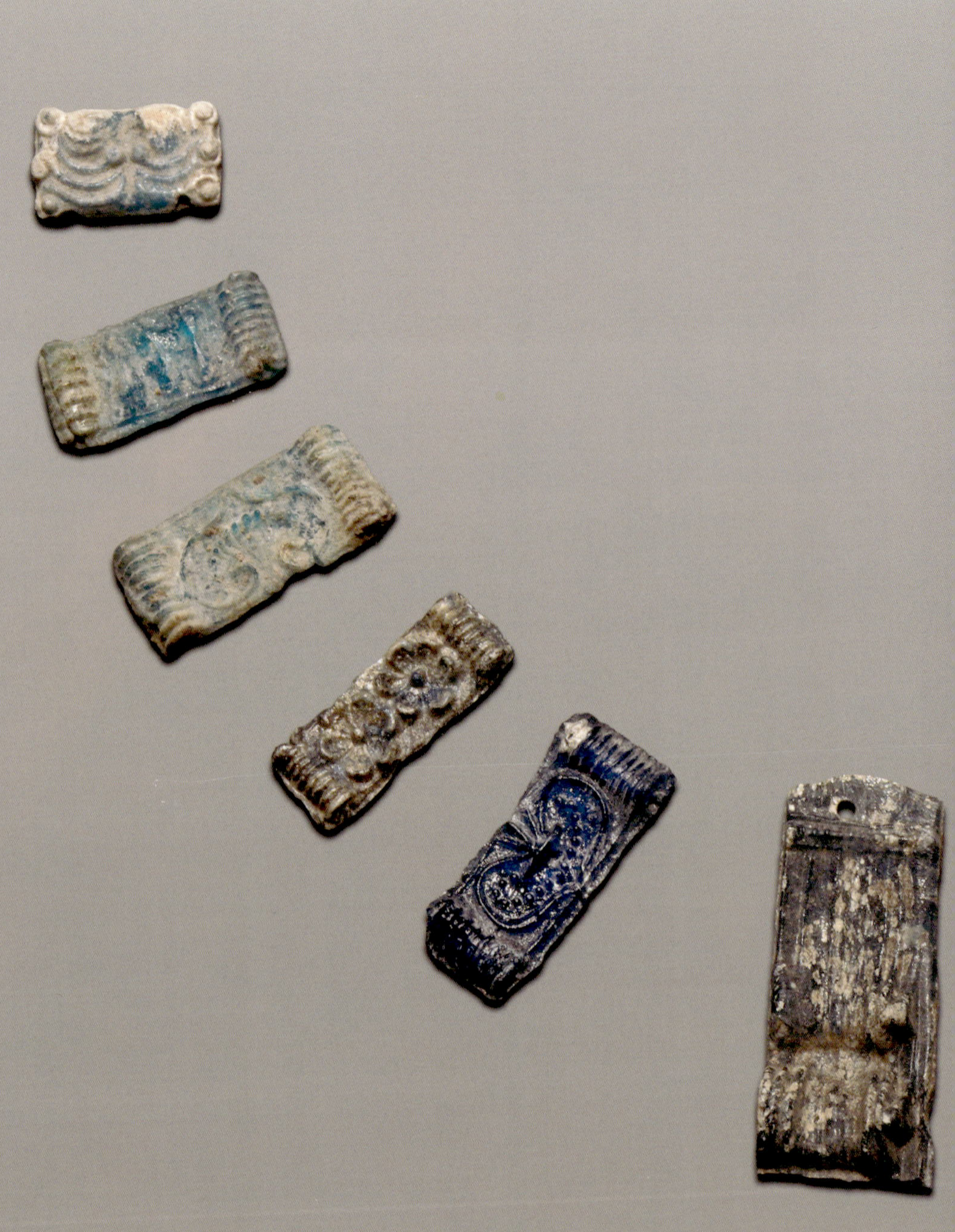

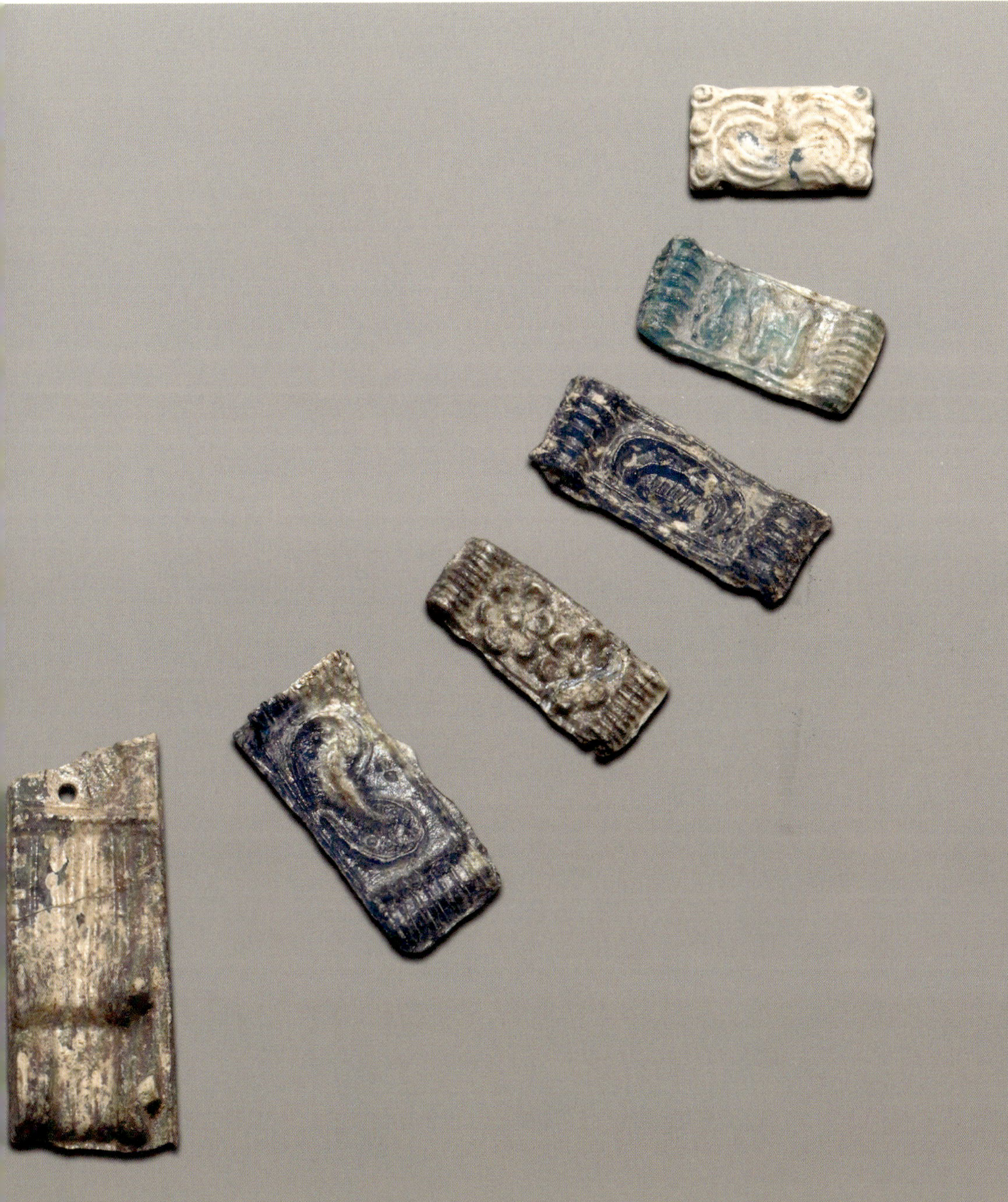

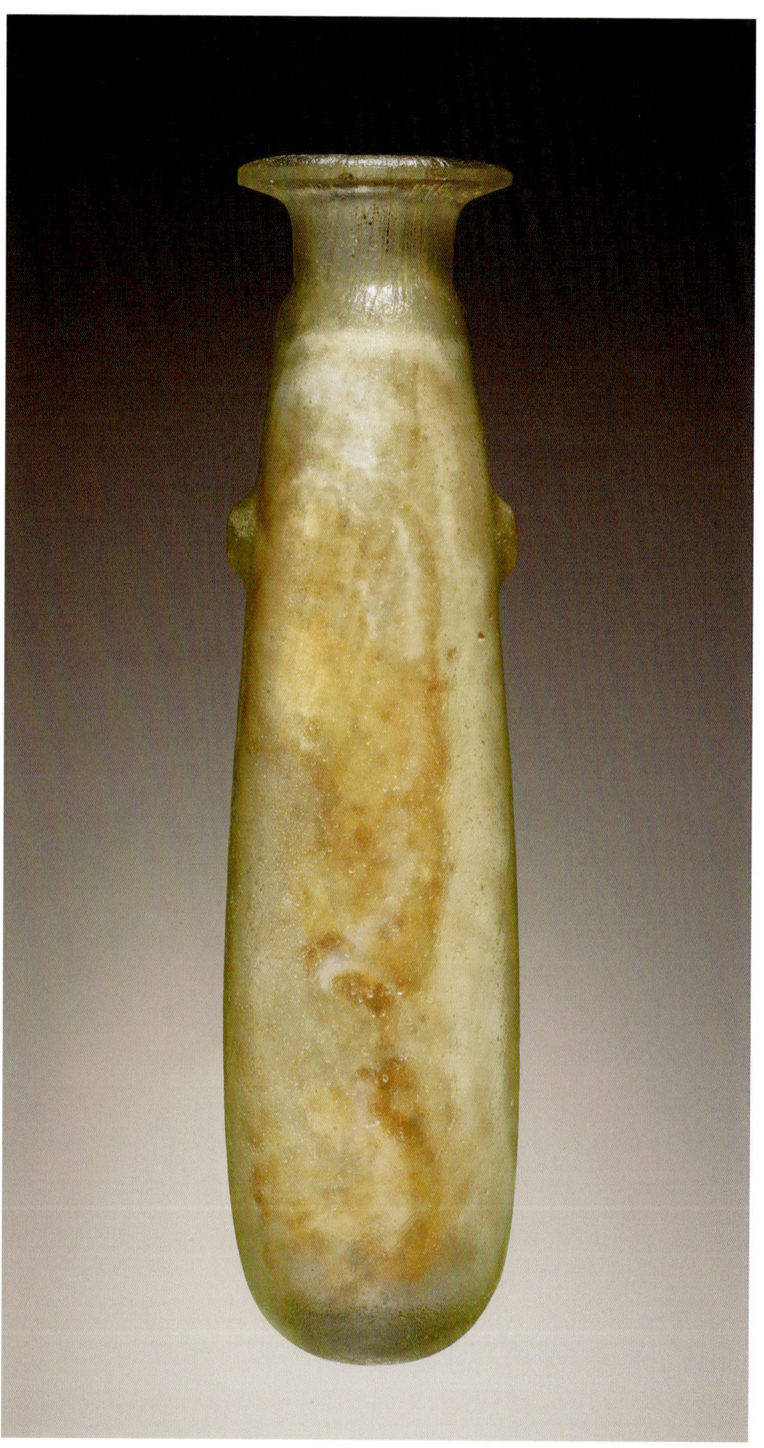

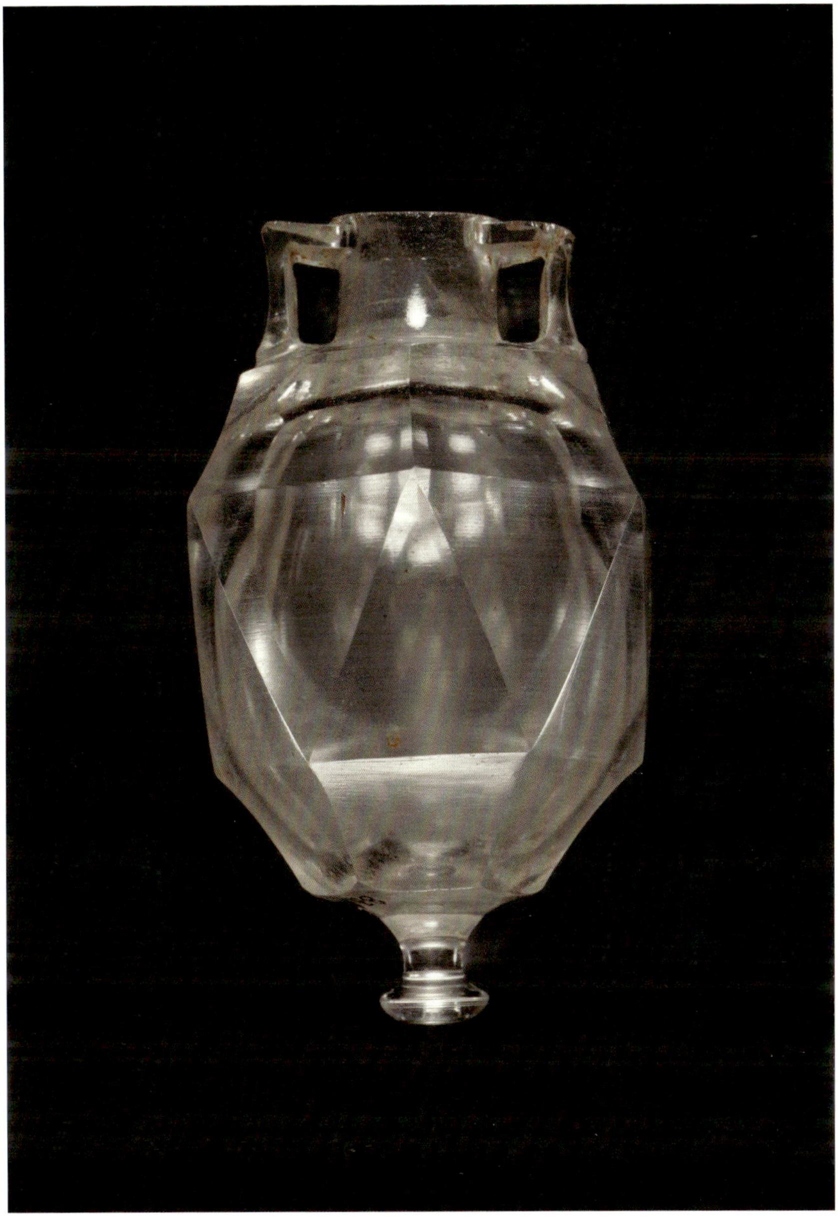

FIGURE 12
Cast perfume flask (alabastron). Assyrian or Achaemenid Persian, seventh–sixth century B.C.
H 14 cm (5½ in.). JPGM 2004.16.

FIGURE 13
Rock crystal flask. Roman, ca. 25 B.C. H 5.7 × DIAM 3.2 cm (2¼ × 1¼ in.).
JPGM 83.AN.331.

MOSAIC GLASS

Although mosaic glass is most frequently associated with the later Hellenistic and Roman periods of glassmaking, there was an early phase of production in Mesopotamia. In the fifteenth century B.C., a manufacturing technique was developed in which small pieces of colored glass were carefully placed within a mold to create a deliberate pattern. When heated, the pieces fused together. The majority of the surviving vessels made in this technique were small and cylindrical, and the glass pieces were arranged in zigzag patterns that encircled the body. A similar design is almost always found on vessels made using the third early technique of glassmaking, called core forming.

CORE-FORMED GLASS

This technique takes its name from the manufacturing process (Figure 14). The first step in the process was to take a metal rod and place around one end a mixture of material such as clay, mud, or sand with an organic binder like dung. This mixture formed the core around which the hot glass was placed and ultimately served to define the interior cavity of the vessel (1). After the core was shaped, it was covered with glass in some manner. The thick viscosity of the glass hampered any effort to simply dip the core into the crucible of hot glass, so the glass must have been removed and applied over the core (2).

Once the core was covered, the glassmaker then rolled this **gather** of glass onto a smooth surface (probably of stone, now called the **marver**) to even out the glass and achieve uniform thickness and smoothness (3). He then took another metal rod and dipped the tip into hot glass of a contrasting color (4). This secondary glass was used to decorate the exterior of the vessel with stripes of this color, achieved by winding a thin **trail** of hot glass around the outside. This striping could be done multiple times with different colors to achieve a polychromatic effect (5), but each time a new trail was added, the vessel needed to be reheated by inserting it into the glory hole of the glass kiln and remarvered (6). Reheating allowed the vessel to receive the new glass more easily and uniformly and avoided thermal shock that could occur when there was a difference in temperature between the glass of the body and the glass of the trail. After all the desired trails

---

FIGURE 14
Sequence of steps to make a core-formed vessel.

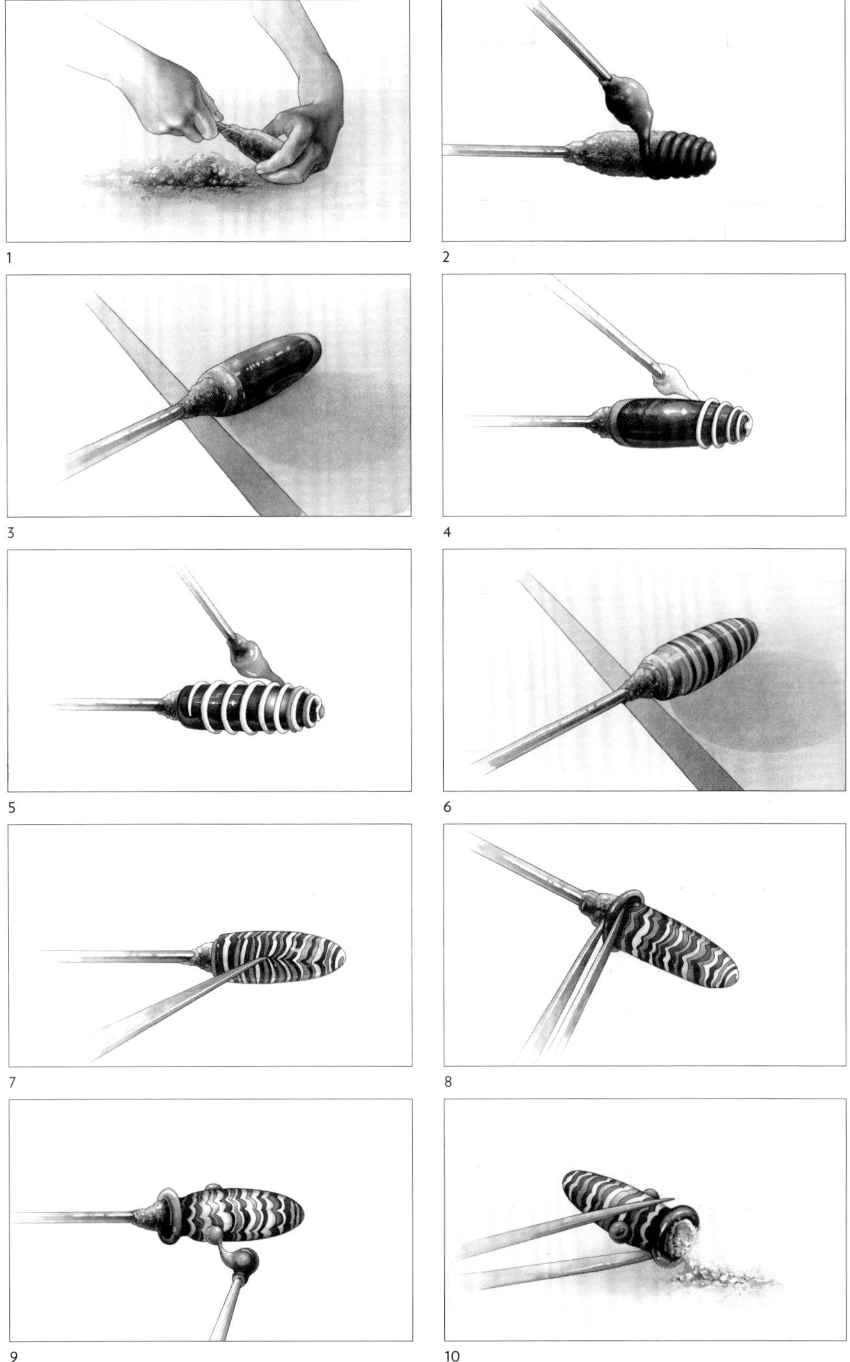

of glass had been added, the glassmaker then took a tool with a pointed end and dragged it up and down the body of the vessel to create a zigzag or wavy pattern (7). When the pattern was complete, the rim was shaped (8), and the handles added (9). Finally, the vessel was rolled again onto the marver to establish a uniform thickness to all the added layers of glass, and then annealed. Once the vessel had cooled, the metal rod and core could be removed (10). Some scholars believe that a separating agent was applied over the core to facilitate its removal from the glass vessel's interior, but traces of such an agent have yet to be detected.

Core forming was used extensively in both Mesopotamia and Egypt for the manufacture of small perfume and cosmetic bottles decorated with multicolored trails (Figure 15). The most frequent color combination found on these vessels is dark blue with yellow or white overlays, but core-formed vessels were also made in imitation of stone, and sometimes white glass was used to replicate alabaster or marble (Figure 16).

The earliest glass vessels made in this manner, dating from the late sixteenth and early fifteenth centuries B.C., have been found at archaeological sites in northern Mesopotamia. Most of these early vessels have been found in the ruins of temples, palatial residences, and graves but rarely in private homes, which has led archaeologists to conclude that they were considered luxury objects and made by a select group of craftsmen for an aristocratic or priestly clientele.

Core-formed glass was also used to enhance architecture, and rods of blue-and-white glass from about 1250–1200 B.C., whose striped patterning resembles candy canes, have been recovered from the Elamite site of the Choga Zanbil ziggurat near Susa (Figure 17). These rods were perhaps placed in some geometric pattern around door and window frames.

These techniques—casting, mosaic, and core forming—were used to make glass objects and vessels for hundreds of years, from the time of the earliest cast wares (small objects and beads) around 2500 B.C. until well into the Hellenistic period in the third and second centuries B.C. We are fortunate that their durability has allowed numerous examples to survive, as it is these surviving objects that have provided art historians with the opportunity to create typologies of shape and design and enabled archaeologists to uncover more sites of glassmaking as the industry finally spread out of the eastern Mediterranean and Egypt in the eighth and seventh centuries B.C. One of the centers of production of core-formed perfume vessels was located on the island of Rhodes beginning in the mid-sixth century B.C. Core-formed vessels were spread far and wide throughout the entire Mediterranean region by Phoenician traders plying the waters with their ships full of merchandise as the ancient world, like our own today, became a much smaller place due to these frequent cultural contacts between the various peoples inhabiting its shores (Figure 18).

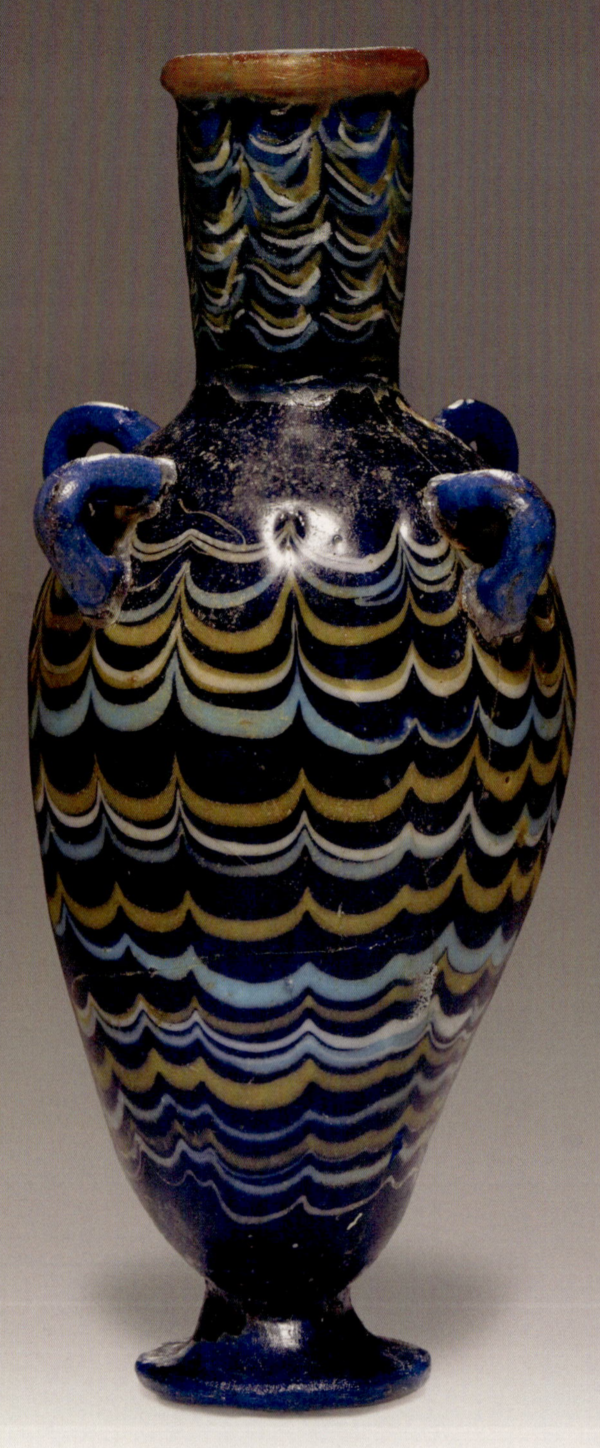

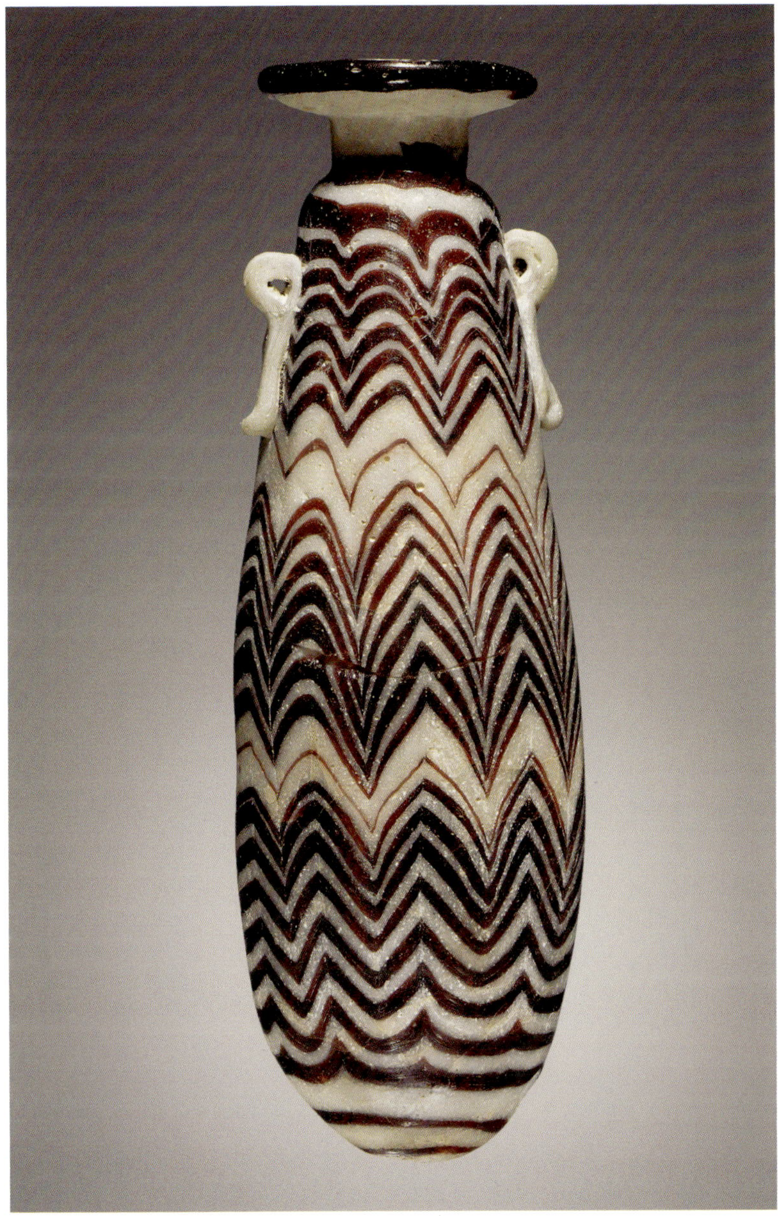

FIGURE 15
Core-formed miniature storage jar (amphoriskos). Egyptian, Eighteenth Dynasty, from the reign of Amenophis III/IV, 1403–1347 B.C. Preserved H 9.5 cm (3¾ in.). As restored: H 12.6 cm (4¹⁵⁄₁₆ in.). JPGM 2003.146.

FIGURE 16
Core-formed perfume flask (alabastron). Greek, sixth–fourth century B.C. H 13.2 cm (5³⁄₁₆ in.). JPGM 2004.5.

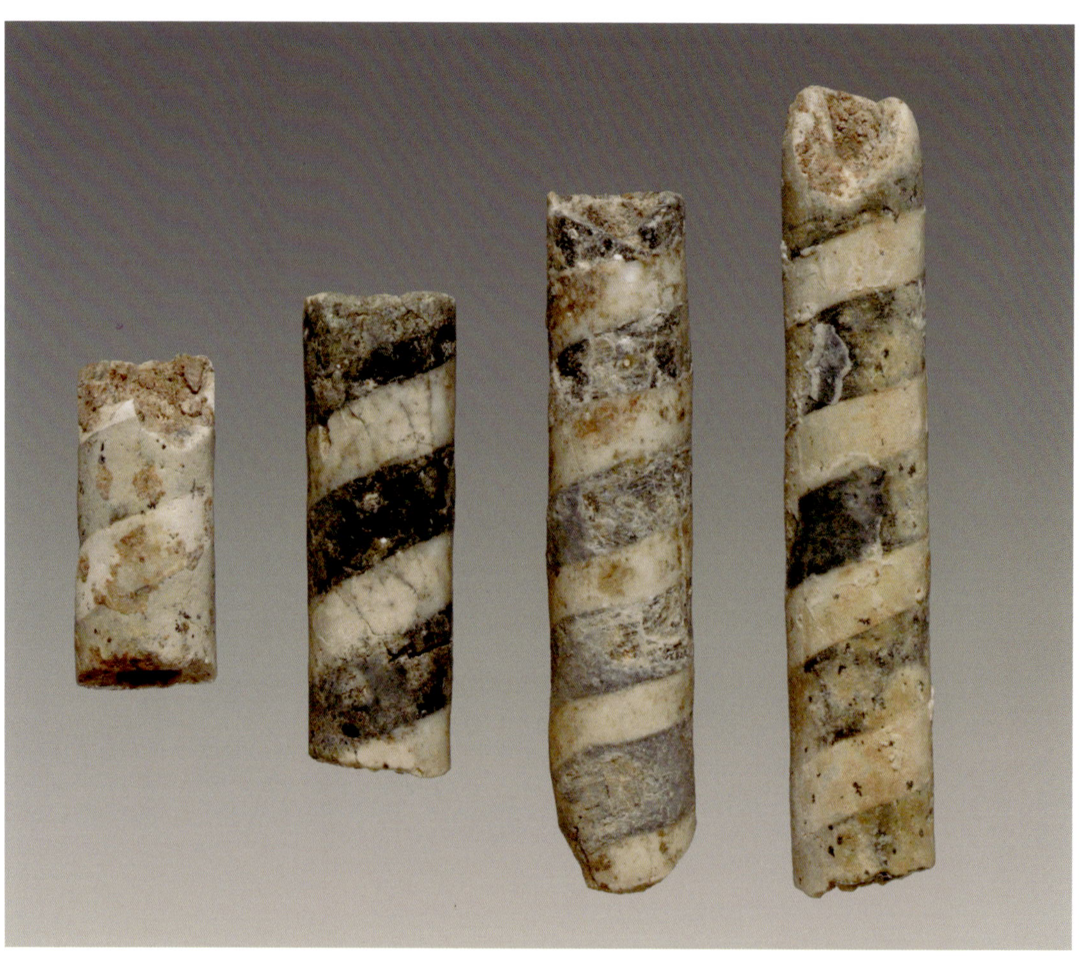

FIGURE 17
Four fragmentary core-formed architectural rods from the site of Choga Zanbil. Elamite,
1250–1200 B.C. L 3.3 × DIAM 1.6 cm (1⁵⁄₁₆ × ⁵⁄₈ in.); L 4.8 × DIAM 1.6 cm (1¹⁵⁄₁₆ × ⁵⁄₈ in.);
L 7.2 × DIAM 1.6 cm (2¹³⁄₁₆ × ⁵⁄₈ in.); L 8.3 × DIAM 1.5 cm (3¼ × ⁹⁄₁₆ in.). JPGM 86.AF.522.

FIGURE 18
Core-formed perfume flask (alabastron). Greek, fourth–third century B.C. H 16.5 cm (6½ in.).
JPGM 2003.193.

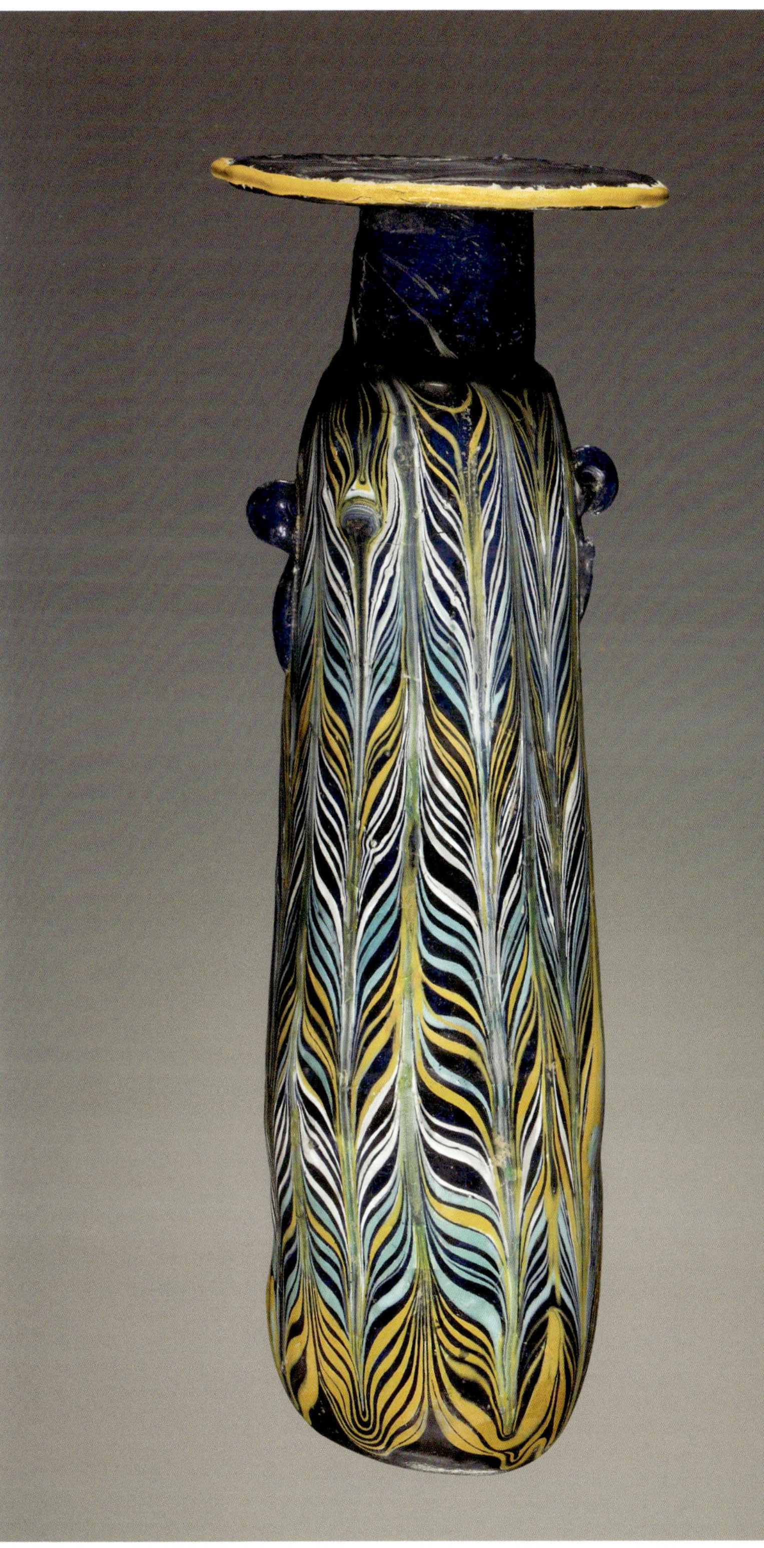

CHAPTER 3

Beginning in the eighth century B.C., glass production started to spread further around the Mediterranean. A small number of Etruscan pieces have been recovered from tombs in Etruria, attesting to glassmaking in the Italian peninsula by the seventh and sixth centuries B.C. (Figure 19). As in earlier periods, the small bottles used for perfumes and cosmetics, small dishes, jewelry pieces, and inlays formed the majority of the glass being produced from the eighth century B.C. on (Figure 20). Its limited and laborious production kept glass firmly in the realm of luxury goods. And by the time of Periclean Athens (fifth century B.C.), its value warranted its inclusion in some of the most important and costly artistic commissions of the age.

We know from both literary descriptions and archaeological remains that glass inlays were used in the construction of one of the Seven Wonders of

Glass in the Mediterranean World

the Ancient World, the large cult statue of Zeus created in about 432 B.C. by the Greek sculptor Pheidias (ca. 490–430 B.C.). The statue stood in the god's temple at Olympia, among the most sacred sites in all of Greece and the birthplace of the Olympic Games (Figure 21). The statue, of a type known as chryselephantine, was made from a variety of materials, including wood, ivory, and also glass. These diverse elements were assembled over an internal armature. The archaeologists who excavated at Olympia uncovered the remains of the workshop of Pheidias in the 1950s, and among the artifacts found were terracotta molds used to cast the colorless glass elements that enhanced the drapery on the statue and its throne.

In the making of the larger drapery pieces, it is likely that a flat pane of glass was heated in a mold that had a wide edge to support the pane as it sagged. Some of the ancient literary sources, such as the travelogue of the Greek geographer Pausanias (*Description of Greece*, 5.1.11.1), describe the drapery as made of gold, but more recently glass scholars have theorized that the gold described was actually gold foil that was placed behind colorless glass inlays to make it more reflective.

We also know that glass continued to be used to enhance architecture. One of the most important architectural projects during the Classical period was the construction of the various temples and sacred buildings on the Acropolis in Athens. Among the most hallowed of these was the Erectheum, where glass was used to decorate the column capitals. Although the glass does not survive today, it was still in place during the seventeenth and eighteenth centuries, and some of the drawings of the ruins on the Acropolis that were made by early European visitors, such as James Stuart and Nicholas Revett, recorded these inlays still in place in the early 1750s.

During the Hellenistic period (323–31 B.C.) there were great strides forward in finding ways to exploit further the manufacturing techniques of glass vessels. The mosaic technique, which far outstripped casting and core forming in the complexity and colors of its patterns, resulted in some of the most spectacular pieces of glass to have survived from antiquity.

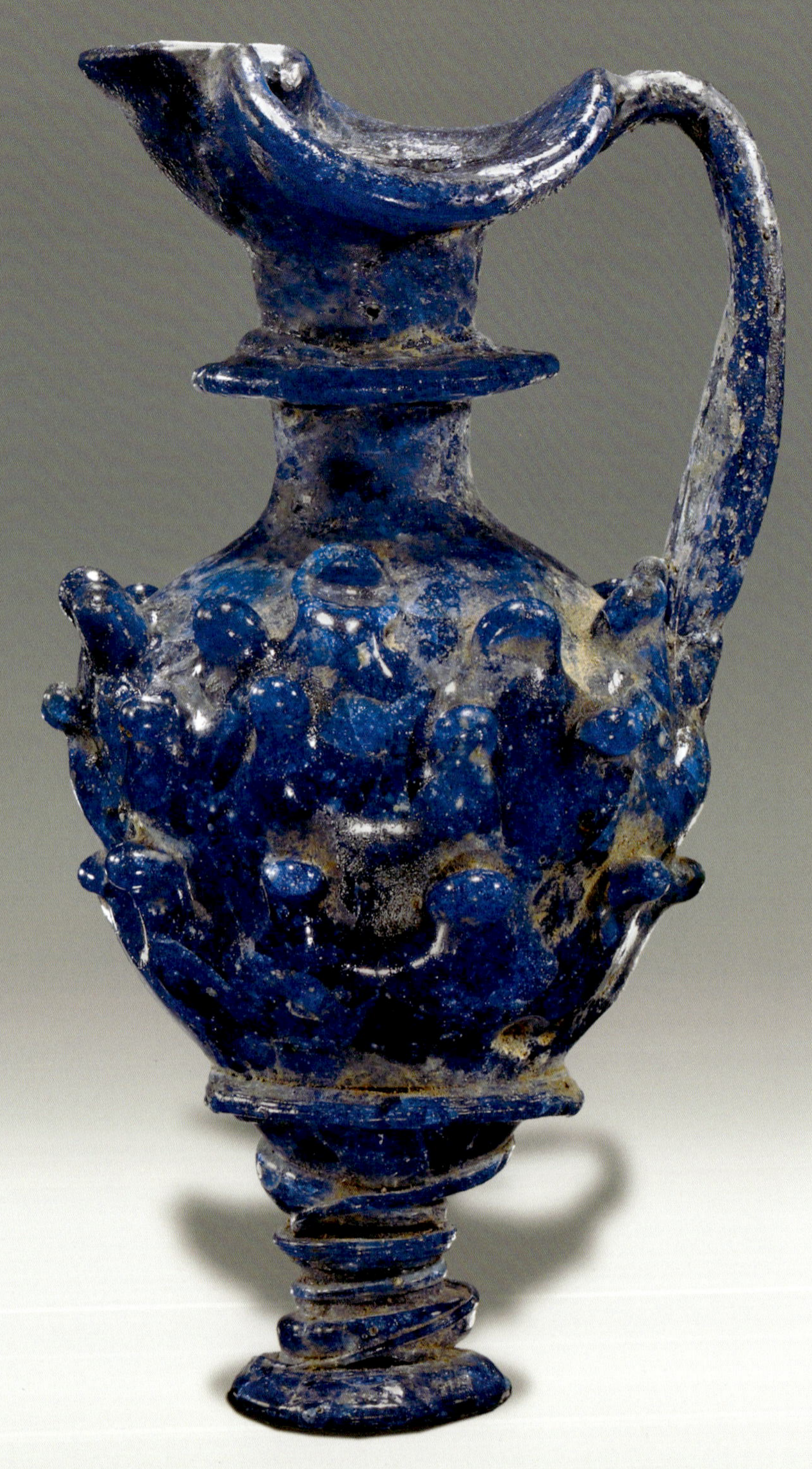

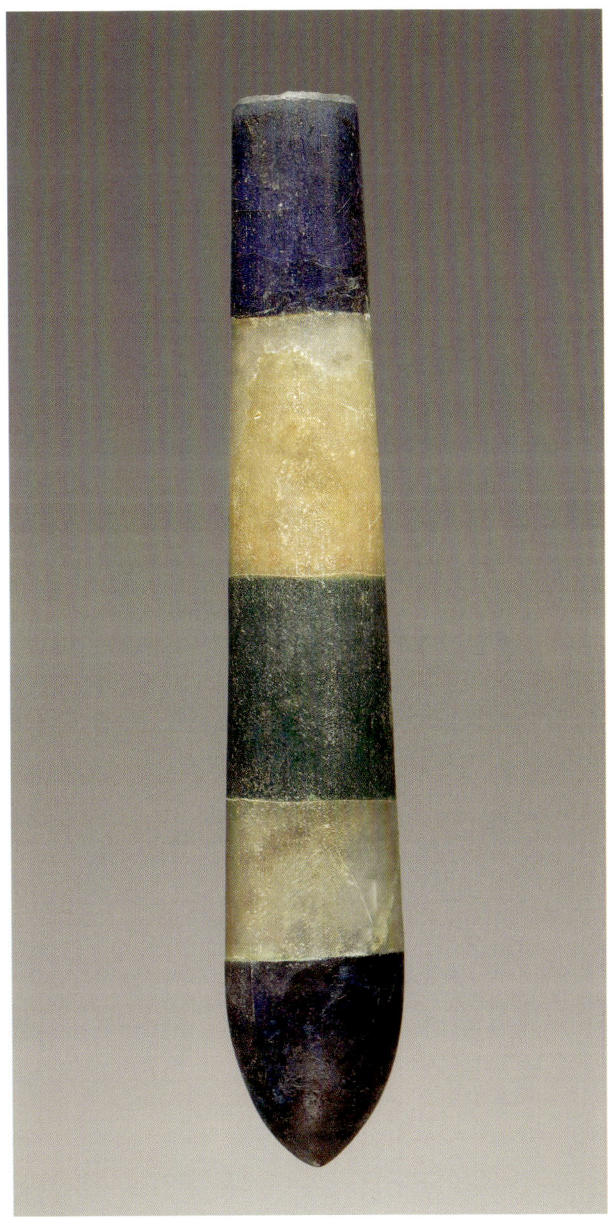

FIGURE 19
Core-formed miniature pitcher (oinochoe). Etruscan, seventh century B.C. H 8.8 cm (3 7/16 in.). JPGM 2003.164.

FIGURE 20
Segmented, core-formed perfume flask (alabastron). Late Hellenistic or Roman, 25 B.C.–A.D. 25. H 13.4 cm (5 1/4 in.). JPGM 2004.21.

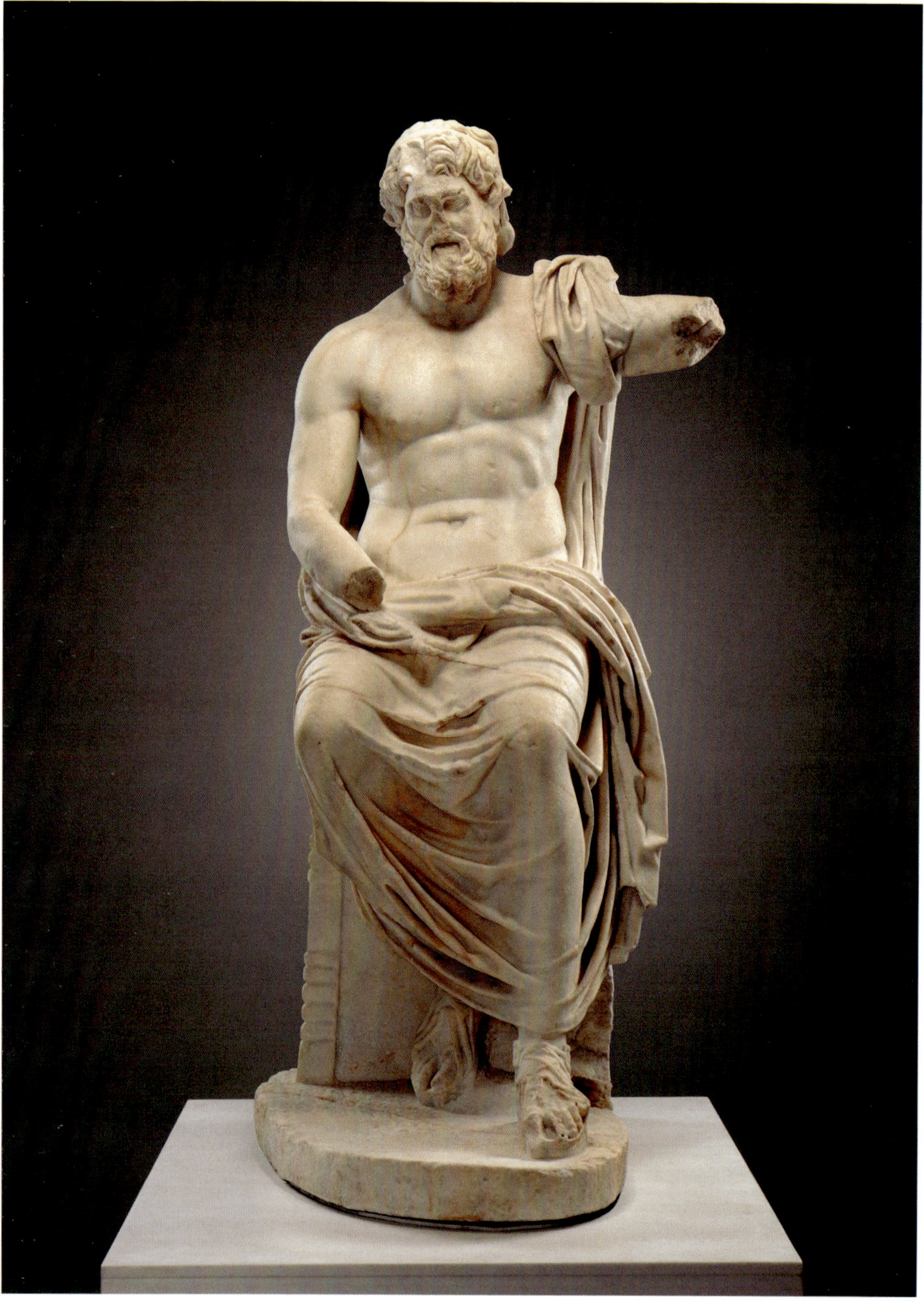

The Marbury Hall Zeus. This statue of Jupiter (Greek Zeus) was found at the villa of Emperor Hadrian near Tivoli, Italy. The seated figure is a Roman reproduction modeled after Pheidias's statue of the god at Olympia. Roman, first century A.D. H 207 cm (81½ in.). JPGM 73.AA.32.

MOSAIC GLASS

The mosaic glass made in antiquity is sometimes incorrectly referred to as *millefiori*. *Millefiori* is a term that should be used to describe only glass made in Venice long after the Roman period, although it employed the same technique. In addition, mosaic glass should not be confused with the mosaics used for architectural decoration, as the two are quite different. It is called mosaic glass because, like a mosaic floor or ceiling, it is composed of multiple pieces combined together.

As with core forming, making a mosaic glass vessel involved several steps (Figure 22). The first focused on the formation of the small pieces of glass that would be combined to form the pattern. These pieces could be monochromatic (as were their earlier Mesopotamian counterparts) or patterned. If they were patterned, then the rods, or **canes**, designed with these patterns had to be made first (1–8). Many canes were designed with multiple colors containing spiral patterns, central dots, floral patterns, or even figural elements. These designs were achieved by carefully working with layers of glass of different colors that were reheated and pushed into shape. Modern glassmakers who design such canes today end up with a square or round block of glass that is about 2 or 3 inches high by 2 or 3 inches wide, but 3 or 4 inches deep. These patterns can be miniaturized by heating the block and attaching solid rods (pontils) at either end. As the pontils are pulled in opposite directions, they stretch the glass like taffy, and the design is miniaturized as the diameter decreases (9, 10). Once the resulting cane reaches the proper dimension, it is removed from the two pontils and annealed. Depending upon the length of the cane, it can be left as is or broken into shorter lengths (11). After the cane has cooled, it can be cut into segments for use in the construction of a mosaic vessel (12).

As with their Mesopotamian precursors, Hellenistic and Roman mosaic vessels were created by placing the segments of glass side by side. Some modern glassmakers who have attempted to replicate this technique believe that the majority of these works were made by arranging the cane segments on a flat surface in order to better control the pattern making (13, 14). Once the glass had been heated and fused into a disk, another cane could be wound around the outside to form the rim of the vessel it would become (15). The disk was then sagged over, or into, a mold that imparted the shape of the vessel to the glass (16). It is possible that this technique, as well as the one described earlier in which the segments are

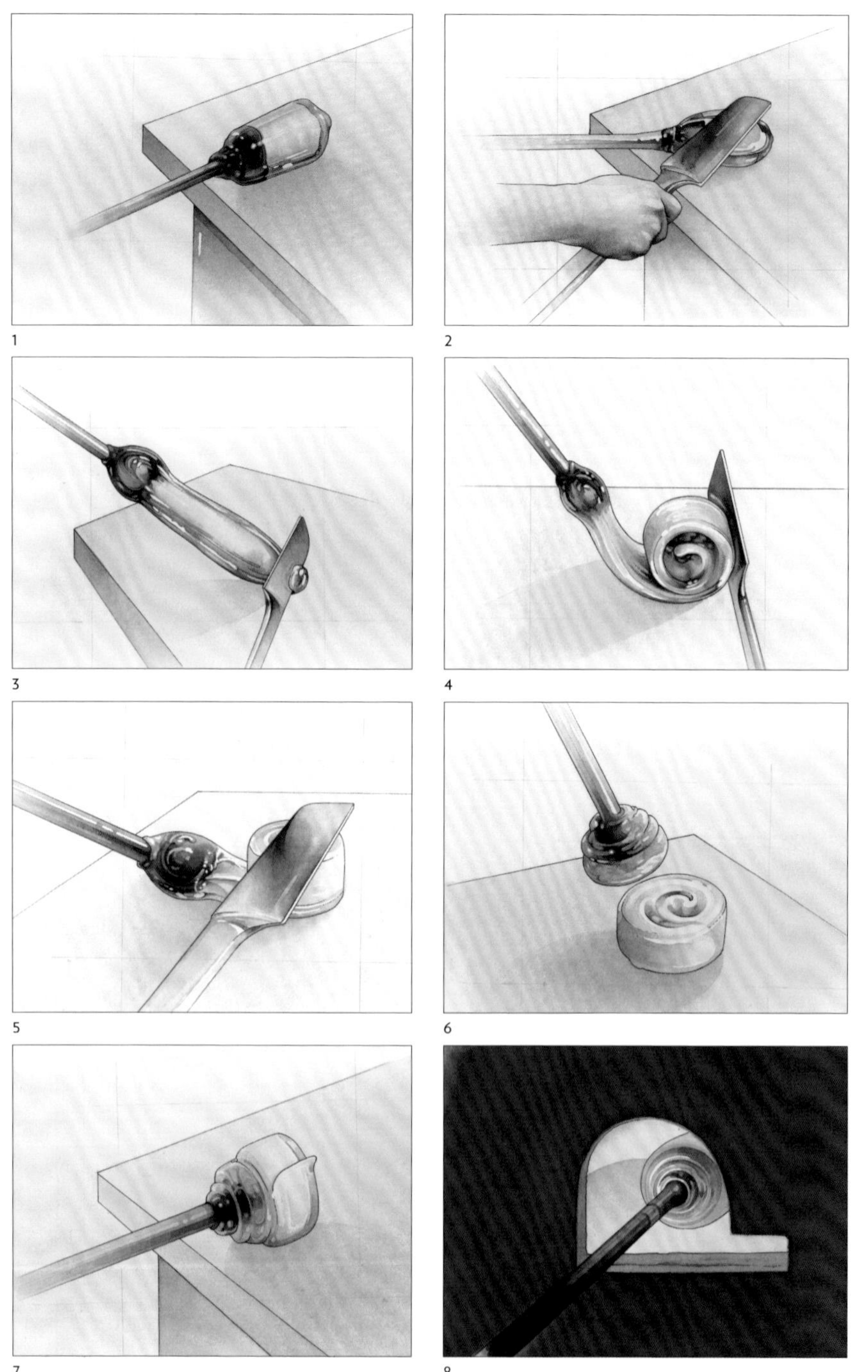

FIGURE 22
Sequence of steps to make a mosaic glass vessel using a cane patterned with a spiral.

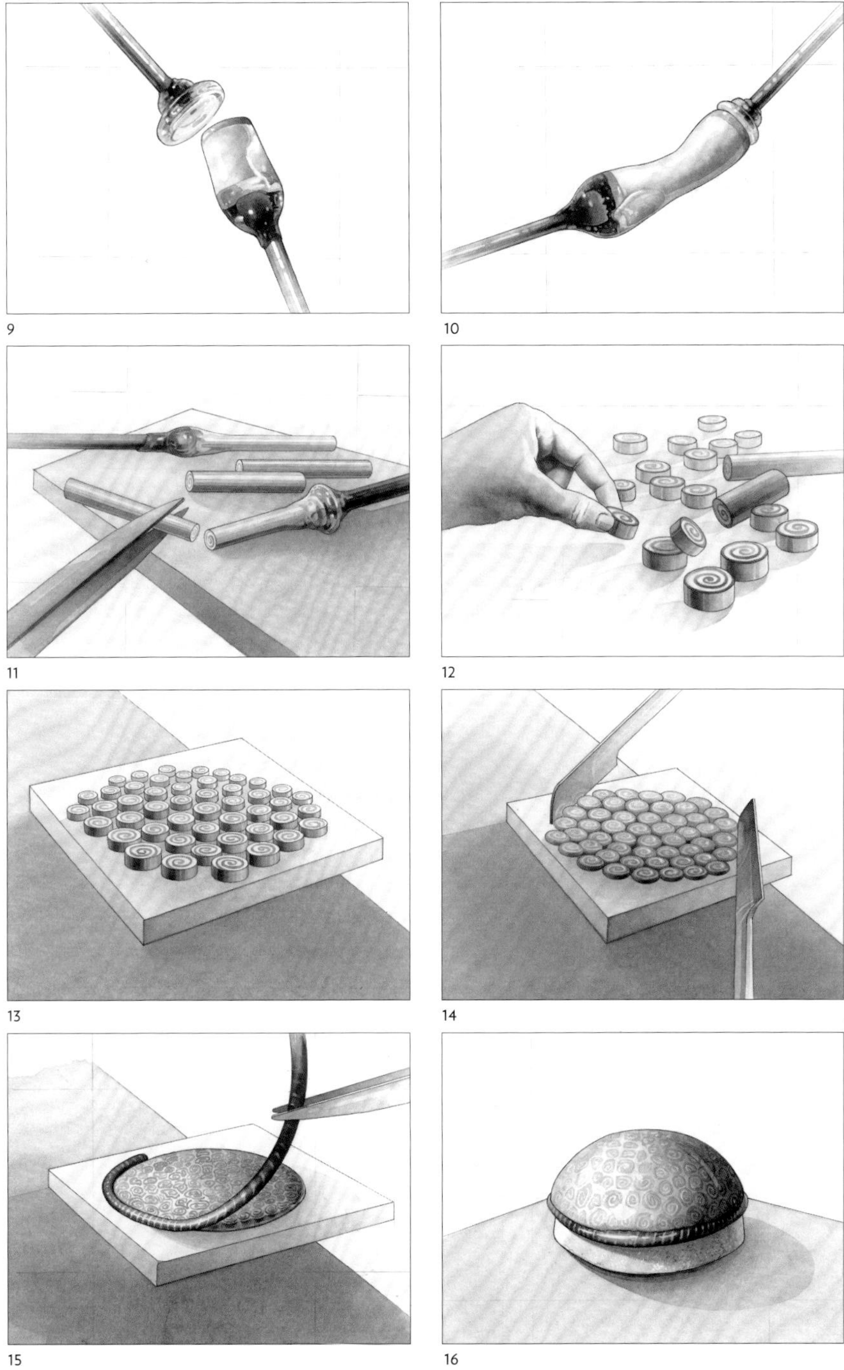

9

10

11

12

13

14

15

16

placed into a shaped mold with an inner mold to hold them in place, were both used in antiquity (see Figure 7, 2b$_2$). Surviving examples include vessels where the cane design is extremely well preserved and easy to understand and which might have been made with a two-part mold, but in others the pattern has heated and diffused, resulting in some truly spectacular patterns but completely obscuring the original cane design, something that could have resulted from glass slumping over a form (Figure 23).

The mosaic glass technique was used primarily to make open bowls and other vessel shapes (Figure 24). The same process used to make the canes for mosaic vessels was also utilized to make flat plaques for jewelry and inlays, and the patterns grew to be extremely complex (Figure 25). Heads of satyrs and actors' theatrical masks were very popular and were made in Egypt during the Ptolemaic period (323–30 B.C.). Floral-patterned pieces also were designed, which, when strung together side by side, made beautiful necklaces.

One innovation of the Hellenistic glassmakers was to create mosaic glass objects that not only used the cut segments of a glass cane but also were patterned with longer cane pieces that were then molded into vessel shapes. This variety of decorative glass is usually called ribbon glass (Figures 26–28). In some instances gold was introduced into the cane design to create an extremely luxurious vessel (Figure 29).

While many of the ribbon glass vessels are extremely colorful, glassmakers continued to try to replicate nature by combining colors and patterns that imitated naturally banded stone such as agate and sardonyx (Figures 30, 31). The glassmakers were extremely effective in their efforts and went to this extreme to imitate natural materials because glass was less expensive and less laborious to produce, and the end product could afford to be bigger than its precious stone counterpart.

But even more challenging than mosaic glass was gold glass, where two vessels, usually bowls of nearly identical size, are nested together and fused, trapping a pattern of gold-foil decoration between them. The process used to make gold glass is complicated and fraught with difficulty. The first challenge was to make two bowls of nearly the same size. This was achieved by casting and grinding them until they made a nesting pair. The artist then applied a design cut from gold foil to the exterior of the inner bowl, sandwiching the design between the two paired bowls, which were then reheated and fused together. How the artist managed to create such objects without overheating the glass and ruining the gold leaf is still being investigated by modern glass scholars, but the successful manufacture of gold glass vessels seems to indicate that the ancient glassmakers had firm control over their production methods.

In addition to drinking bowls, gold sandwich glass also was used for jewelry. A double-sided pendant in the collection of the Getty Museum is made from two oval pieces of clear glass with a gold-foil figure placed on its flat underside. Sandwiched between the two pieces is a ceramic disk that provides a background color for the gold-foil figure of Nike, the goddess of victory, on one side and the figure of a hippocamp, a sort of sea dragon, on the other (Figures 32a, b). In addition to vessels and jewelry, gold glass inlays similar to this gem were used to decorate elaborate pieces of furniture, some of which has been found in Hellenistic tombs in Macedonia.

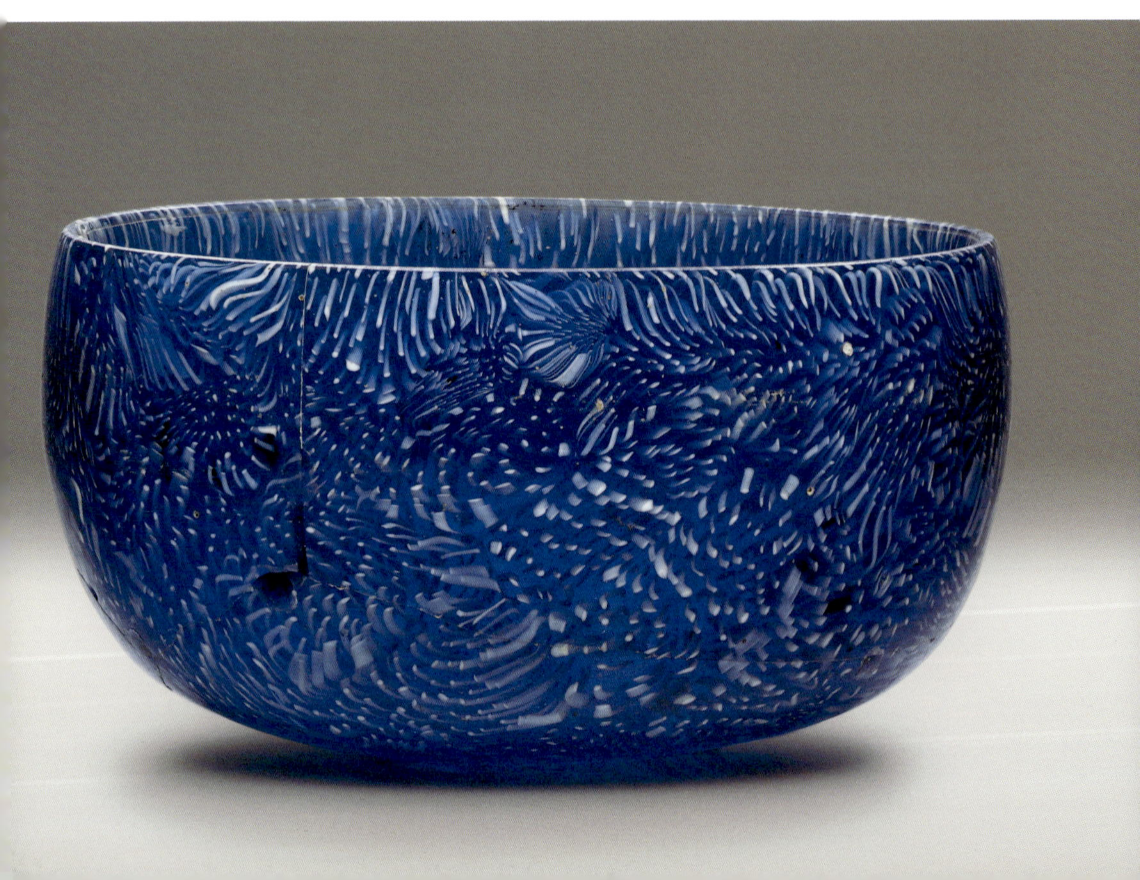

FIGURE 23
Hemispherical mosaic bowl. Late Hellenistic or Roman, first century B.C.
H 5.2 × DIAM 10.2 cm (2¹⁄₁₆ × 4 in.). JPGM 2004.24.

FIGURE 24
Mosaic bowl. Late Hellenistic, first century B.C. H 3.8 × DIAM 10.3 cm
(1¹⁄₂ × 4¹⁄₁₆ in.). JPGM 2003.246.

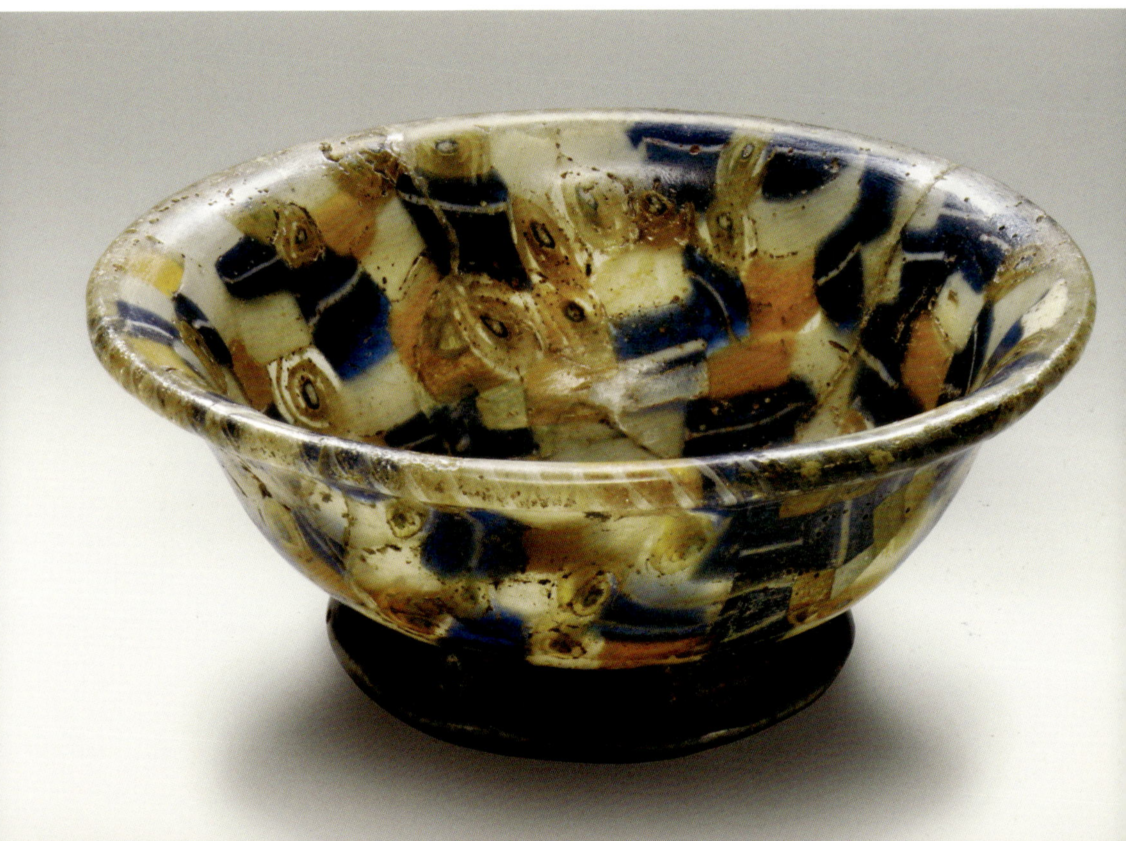

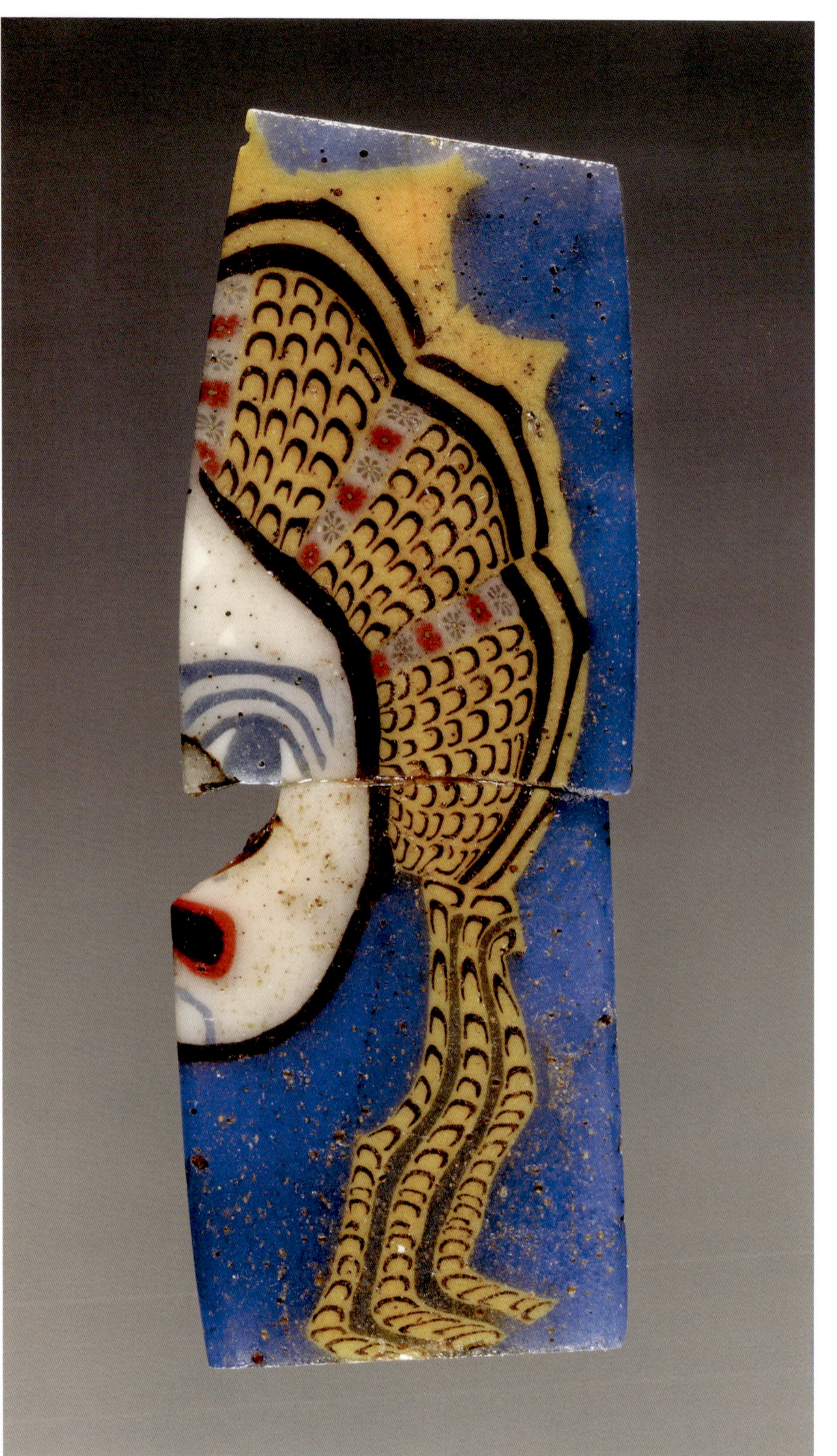

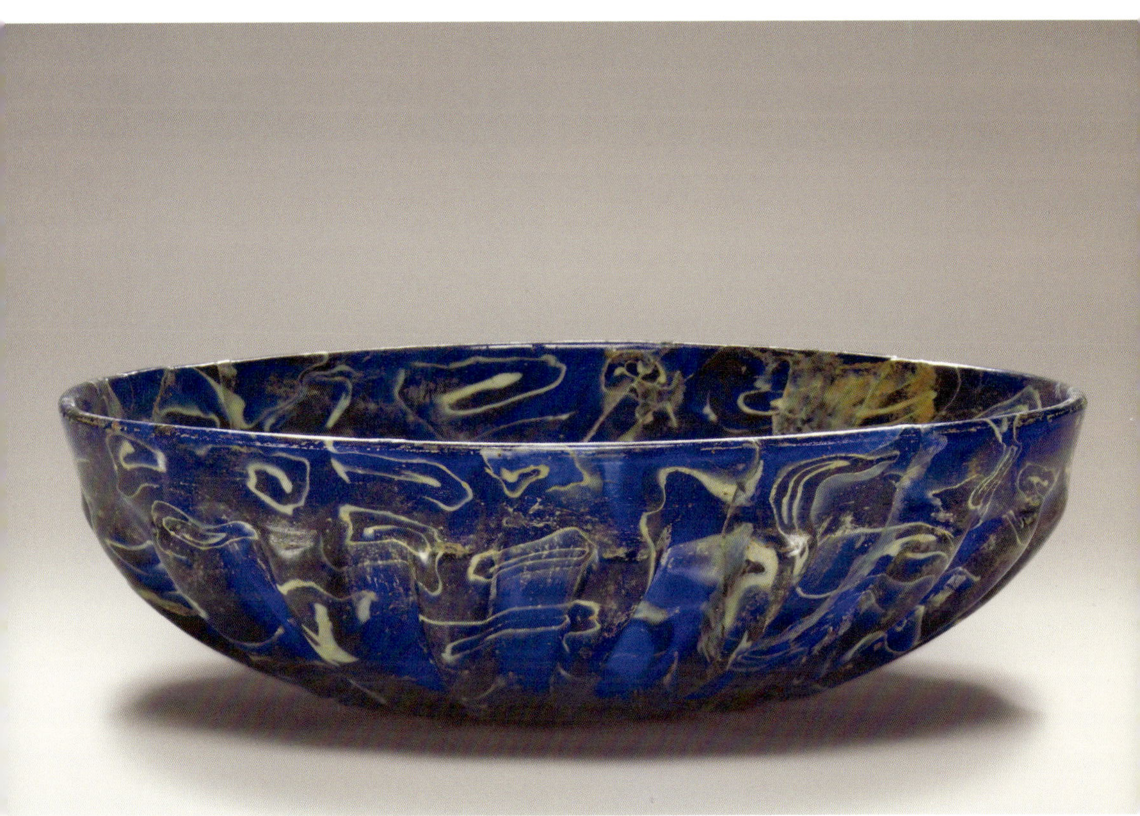

FIGURE 25
Cane slice with blonde-wigged theatrical mask. Late Hellenistic or Roman, first century B.C.–
first century A.D. H 3.4 cm (1⁵⁄₁₆ in.). JPGM 2004.28.

FIGURE 26
Ribbed mosaic bowl. Roman, first century B.C. H 4.9 × DIAM 17 cm (1¹⁵⁄₁₆ × 6¹¹⁄₁₆ in.).
JPGM 2004.25.

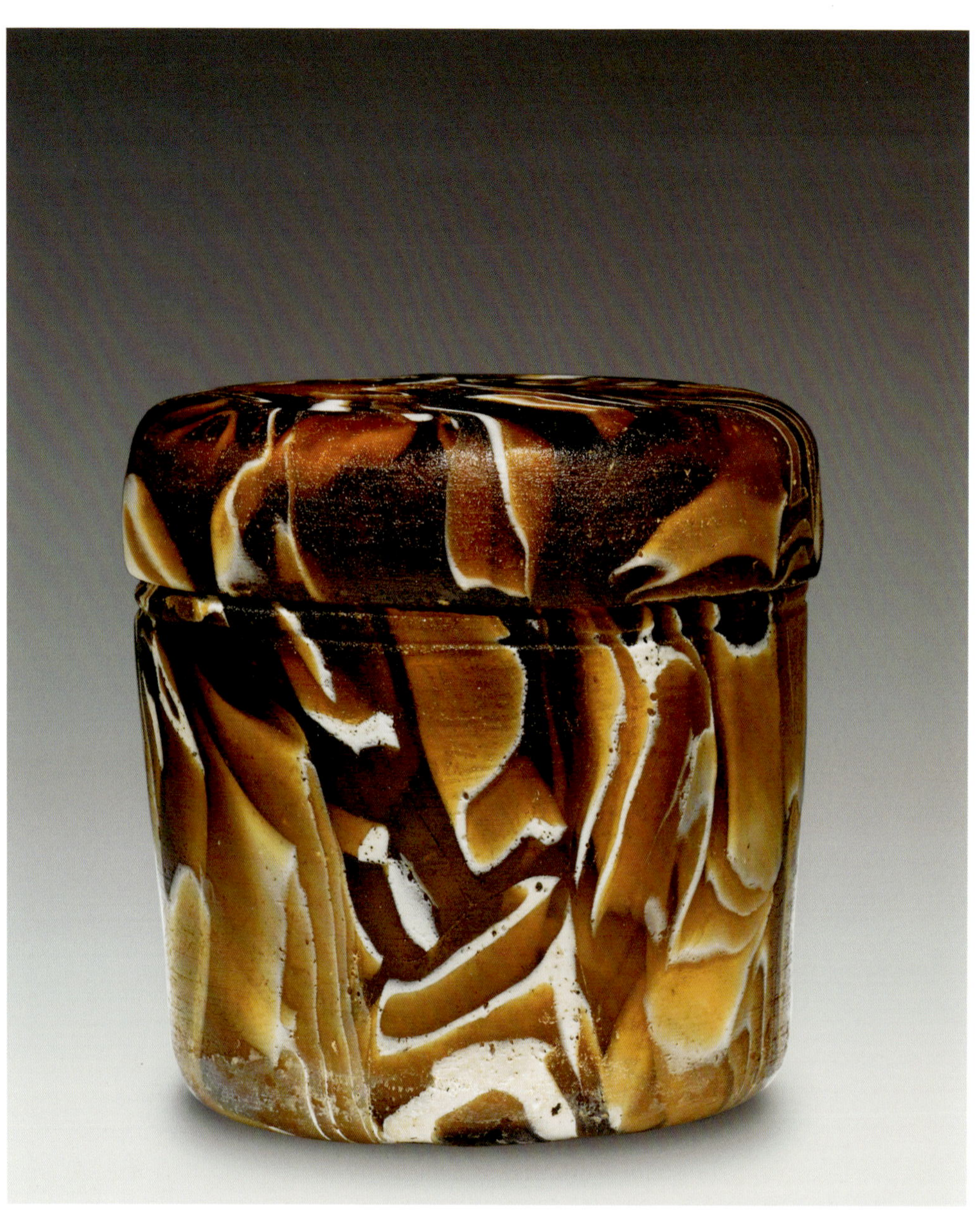

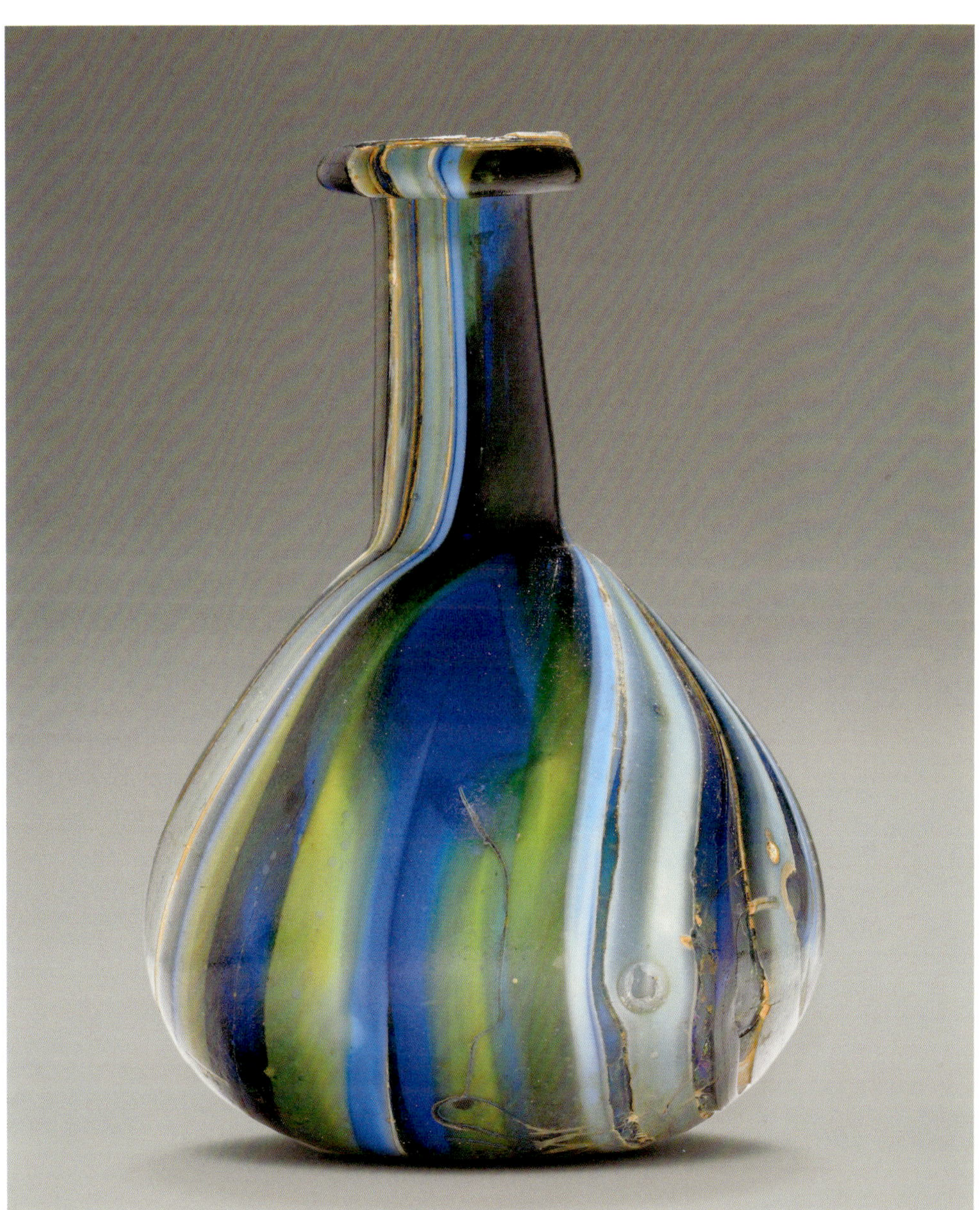

FIGURE 27
Mosaic container (pyxis) with lid. Roman, first century A.D. With lid: H 5.1 × DIAM 5.7 cm
(2 × 2¼ in.). JPGM 2003.256.

FIGURE 28
Ribbon flask. Roman, first century A.D. H 9 cm (3⁹⁄₁₆ in.). JPGM 2003.276.

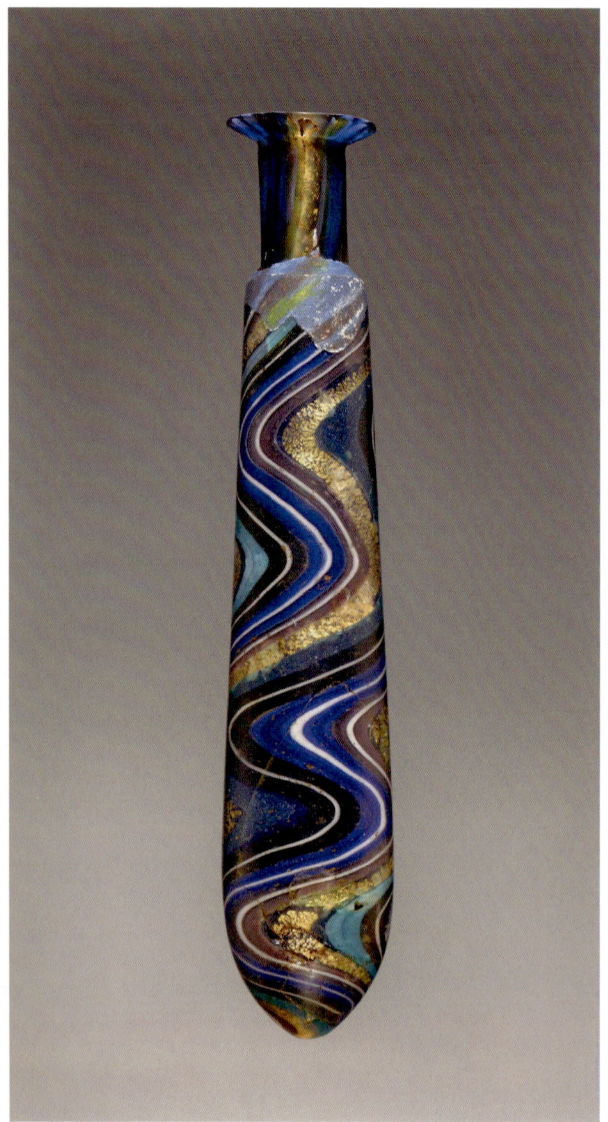

FIGURE 29
Gold band perfume flask (alabastron). Late Hellenistic or Roman, 25 B.C.–A.D. 25. H 13 cm (5⅛ in.). JPGM 2003.229.

FIGURE 30
Ribbed mosaic bowl. Roman, first century B.C.–first century A.D. H 7.5 × DIAM 18 cm (2^{15}⁄$_{16}$ × 7⅟₁₆ in.). JPGM 72.AF.37.

FIGURE 31
Agate bowl. Roman, first–second century A.D. H 3.8 × DIAM 9.1 cm (1½ × 3⁹⁄₁₆ in.). JPGM 72.AI.38.

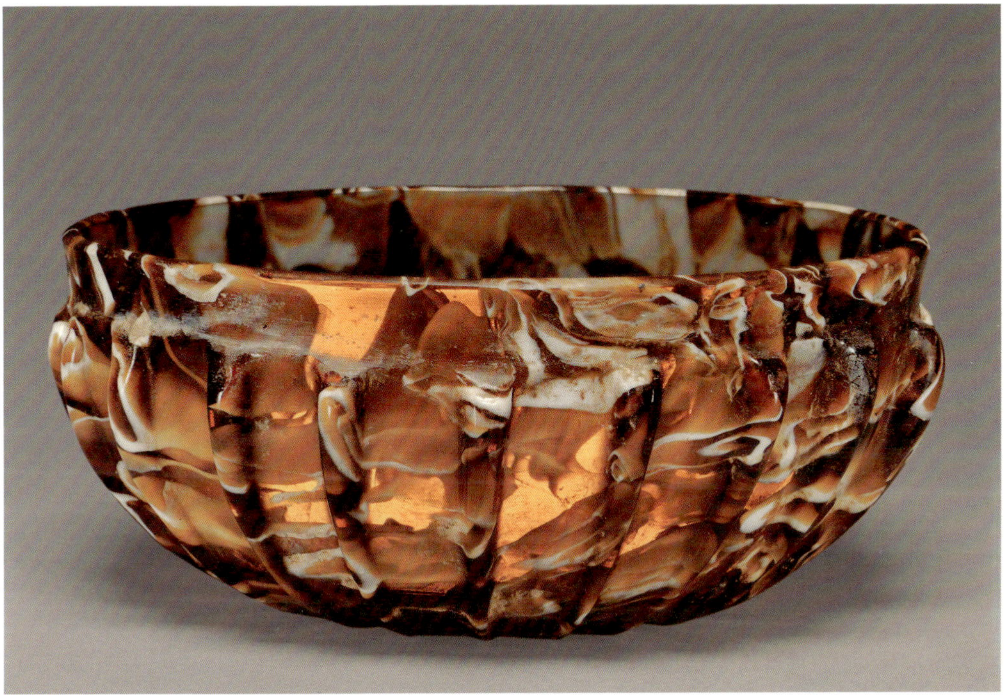

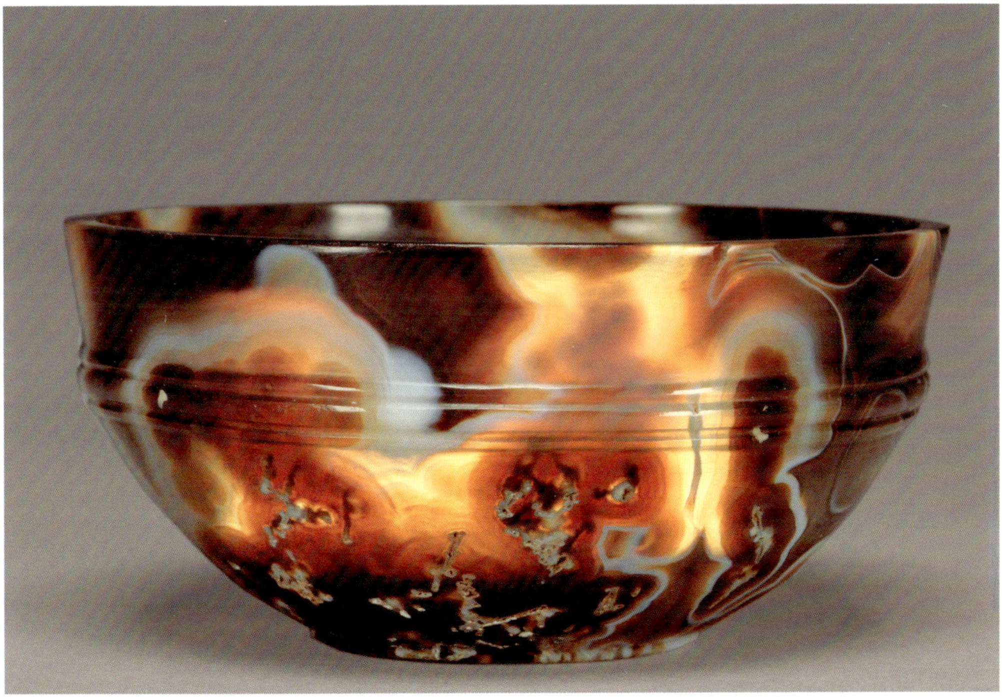

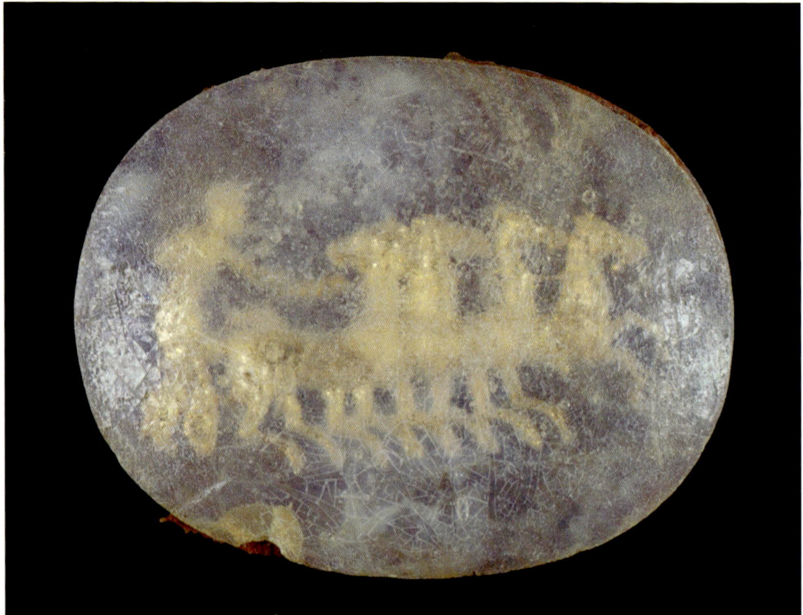

32A.

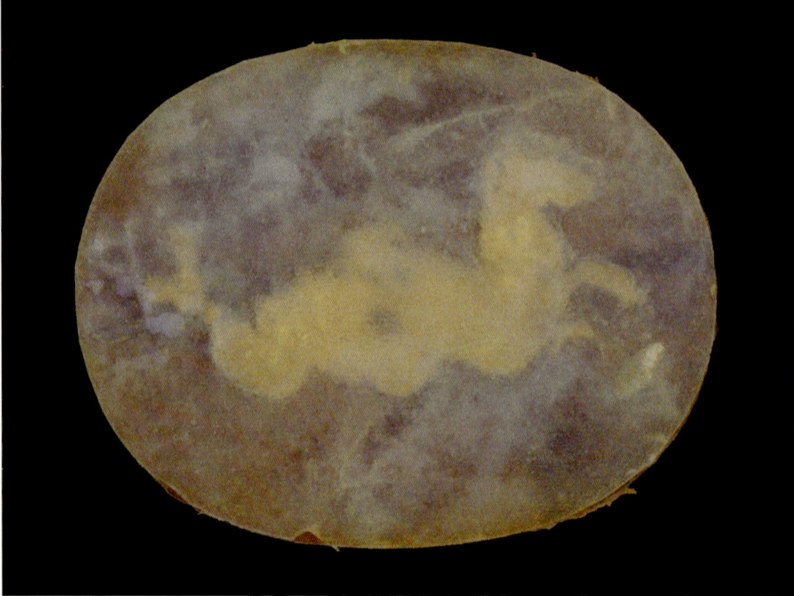

32B.

FIGURES 32 A, B
Double-sided gold glass gem. **A**. Nike, goddess of victory, drives a four-horse chariot.
B. Hippocamp (a sea dragon). Greek, from South Italy, fourth century B.C. H 2.4 × W 1.9 × D 0.8 cm
($^{15}/_{16}$ × $^3/_4$ × $^5/_{16}$ in.). JPGM 88.AN.106.

CHAPTER 4

The technical development that allowed the glass industry to expand enormously during the first century A.D. was the invention of glassblowing. This technique of inflating a bubble of glass at the end of a hollow tube enabled the glassmaker to produce vessels more quickly and less expensively. As we have seen, prior to the advent of glassblowing, each vessel had to be made through a laborious process such as mosaic, casting, or core forming. The quickness of the process of inflation completely changed this and made glass a more affordable commodity. The average citizen could now buy what had previously been seen as beyond his reach. It is at this time that archaeologists have noted a reduction in the amount of ceramic wares being used in households and an increase in the amount of glass being found.

The Discovery of Inflating Glass

The Greek historian and geographer Strabo (63/64 B.C.–ca. A.D. 24) commented that "a drinking cup could be bought for a copper [coin]" (*Geography*, 16.2.25).

The earliest evidence for inflated wares comes from Israel. A small blown-glass bottle dating to the late second and first centuries B.C. was found in an excavation of a series of Jewish tomb-caves in Ein Gedi. And in Jerusalem, the waste of a glass workshop was found in a pool in a first-century-B.C. house in the Old Quarter of the city. Among the waste products were glass tubes whose ends were slightly inflated. The tubes had been heated and then pinched together at one end; blowing a puff of air down the tube resulted in the expansion of the hot glass at its end into a bulb. While these tubes served to begin the process, it was the use of a metal blowpipe to inflate glass that made glassblowing a viable technique.

Two small clay oil lamps found in Italy and Croatia that date to the second half of the first century A.D. provide the earliest known images of a Roman glassmaker sitting in front of his furnace. On the top of the lamps are mold-made scenes in relief that give us our first glimpse of what early glass kilns looked like. The blower sits at the right, his blowpipe held aloft and a vessel inflating on the end. In front of him are two holes in the furnace. The larger one at the bottom was probably the stoke hole, in which fuel was put to keep the furnace hot. The smaller hole above is the glory hole. The artist would insert his blowpipe into the glory hole in order to remove hot glass from the small crucibles resting inside. An attendant crouches at the other side of the kiln and seems to be holding a vessel. He may be removing pieces from the annealing oven. There is a small ledge in front of the glory hole that may have been used for marvering. It is interesting that this marvering ledge is located directly in front of the furnace, because it suggests that the temperatures were low enough for the glassblower to sit in front to work. Some modern glassmakers in Egypt and Turkey working with small furnaces still follow this practice. But as the engineering of glass furnaces progressed through time and the size of the furnaces increased, they became too hot for the blowers to work so close by, and other tools were created, like the chair (see Figure 5).

By the end of the first century B.C., glassblowing was a perfected process, and the metal blowpipe had been developed as a tool. Glassmakers now recognized that the malleable character of glass allowed them to break away from the long-established shapes of vessels—the flasks, bowls, cups, and pitchers found in metal and ceramic—to develop new shapes that were designed, in some instances, to meet specific needs of storage and usage. Narrow-bodied flasks, so-called unguentaria, began to replace the traditional alabastron and aryballos as a preferred shape for holding perfumed oil. Glass bottles began to be used for food storage, as they were easier to clean than ceramic. The transparent quality of glass was recognized for its ability to make food and drink more appealing to diners (Figure 33). As the Stoic philosopher Seneca (Lucius Annaeus Seneca, ca. 4 B.C.–A.D. 65) commented, "Apples seem more beautiful if they are floating in a glass" (*Investigations in Natural Science*, 1.6).

The speed with which the glass industry expanded was due, in part, to the Pax Romana, the period of prosperity and peace established during the rule of

Augustus (23 September, 63 B.C.–19 August, A.D. 14). With the defeat of Cleopatra and Marc Antony at the Battle of Actium in 31 B.C., Augustus controlled an empire that encompassed nearly the entire Mediterranean basin. This consolidation led to freer commerce, and with that, a greater influx of eastern goods, including glass, in the west. While the glass industry had concentrated itself in the eastern part of the Mediterranean until the first century B.C., it now began a migration westward in response to a growing demand for glass, particularly in the new cultural center of the Mediterranean, the capital city of Rome.

The earliest products of the Roman glass industry were those that had been produced before: molded bowls and other cast tablewares (Figures 34–36). Initially, the products of Italian workshops could be differentiated from those of the eastern Mediterranean factories by their use of color, but the vessel shapes themselves were basically the same. What was different in the Roman period, however, was the size of the factories producing them and improved marketing techniques that followed the consolidation of the Roman Empire. Strabo speaks of purchasing glass in Rome itself in the last quarter of the first century B.C. (*Geography*, 16.2.25). A half-century later, Pliny discusses the sand of the River Volturnus, between Liternum and Cumae in Campania, as being excellent for the production of glass (see p. 15) and implies elsewhere in the same passage that glassmaking had spread from Italy into Gaul and Spain. Because this time was a period of peace, the population expanded, and with it came a greater need for all sorts of household goods, increasingly made of glass. It is estimated that by the middle of the first century B.C., the population of the city of Rome may have reached one million people, many of whom ate and drank from—and used various lotions, unguents, and oils stored in—glass containers.

One of the greatest technical achievements of the early Roman glassmakers was the creation of cameo glass. Like the agate vessels discussed earlier (see Figures 30, 31), Roman cameo glass was created to imitate precious layered stones. After gathering and slightly inflating the glass used for the underlayer, two or more different colored glasses were added in layers and fused together. The top layers were carved after annealing to create a relief design on the surface (Figure 37). The knowledge of chemistry and a deep understanding on the part of Roman glassmakers about the materials they were using is evident in the composition of the glass used for cameo glass vessels, for the white overlay contains a greater amount of lead than the underlayer. The higher lead content made the glass both white and softer and thus easier to carve without destroying it. It must have taken much trial and error to determine which opaque white glass recipe was best for carving, and collaboration with the carvers of hard stones to achieve such spectacular results. The most frequently used color combination of opaque white over a dark, cobalt blue was not something found in nature, thus glassmakers imaginatively used the unique qualities of glass and their own ability to control its color to create the new art form of cameo glass. Scholars currently believe that the center for the production of Roman cameo glass was Rome itself.

Some of the largest and best preserved examples have been found in the ruins of houses in Pompeii that were buried by the eruption of Mt. Vesuvius in A.D. 79. Two large plaques with scenes related to the worship of Bacchus, the Roman god of wine, have been found, as well as a large amphora, no doubt used to store wine since it is largely decorated with intertwined grapevines laden with grape clusters, among which scamper small figures of Erotes who are harvesting the grapes.

In Rome, uncarved fragments of cameo glass have been found, which strongly suggests that a workshop responsible for making the blanks for carving was located there. A small perfume flask in the collection of the Getty Museum strengthens that argument, as its decoration has very enigmatic figures, including a striding pharaoh, an obelisk, and a statue of the god Thoth in animal form as a seated baboon (Figure 38). Initially scholars believed that this piece must have been made in Alexandria, because these iconographic elements are Egyptian in style. But monuments and figures, such as those found on the flask, were also present in the city of Rome; Augustus had brought two Egyptian obelisks to the city in 10 B.C., and statues of pharaohs and Thoth as a baboon were present in the sanctuary of the Egyptian goddess Isis. This flask is perhaps a perfect embodiment of Rome in the closing years of the first century B.C., a city at the center of a vast multicultural empire, whose citizens and inhabitants were aware of each others' customs and religious practices, and whose luxury arts were at their height.

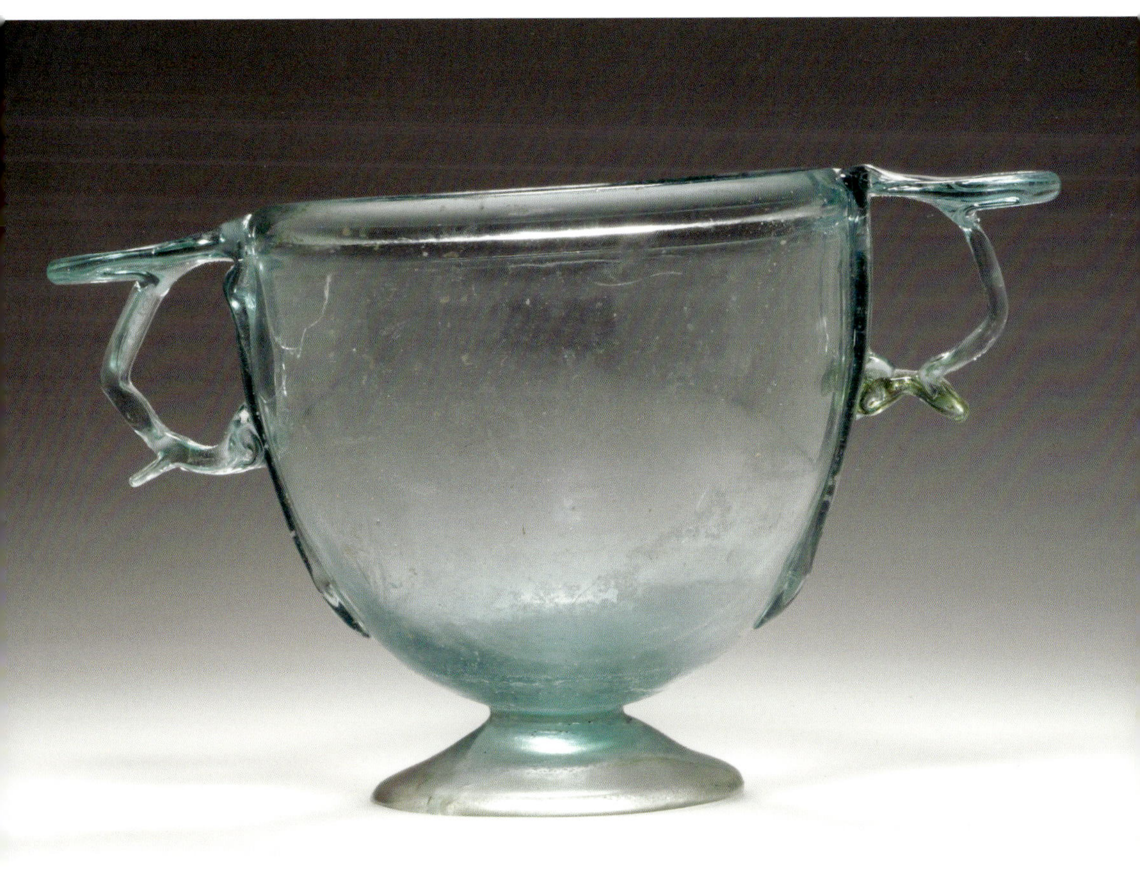

FIGURE 33
Two-handled wine cup (skyphos). Roman, first century A.D. H 8.3 × DIAM 8.5 cm
(3¼ × 3⅜ in.). JPGM 2003.361.

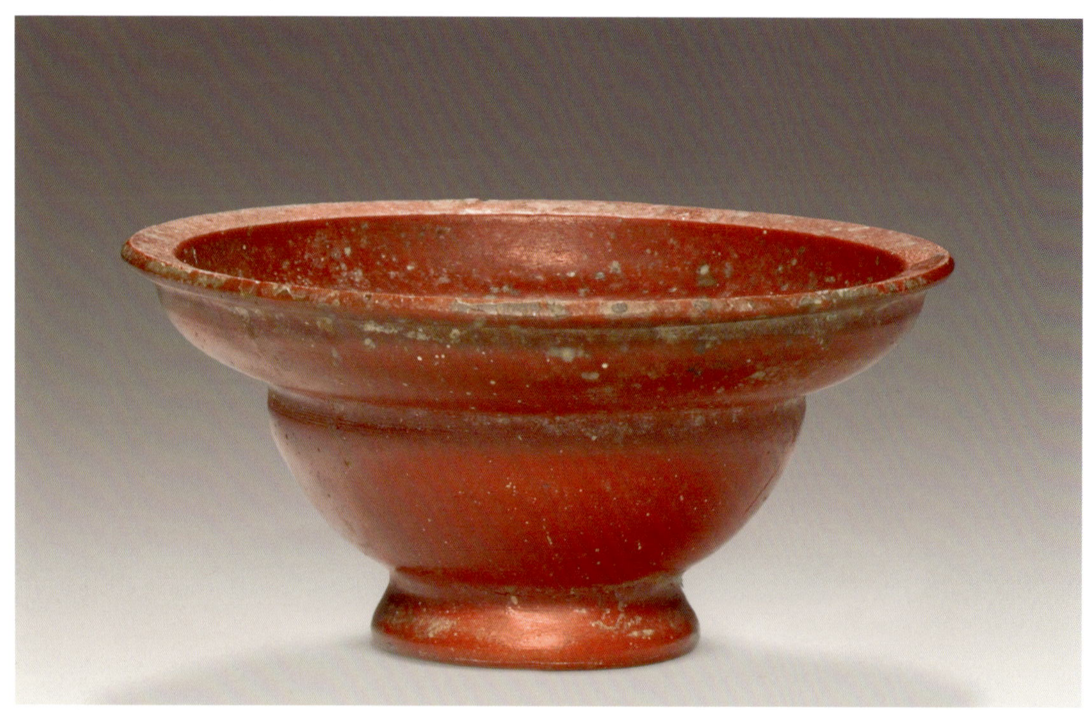

FIGURE 34
Small cast bowl (patella cup). Late Hellenistic or Roman, 25 B.C.–A.D. 25.
H 3.4 × DIAM 7.3 cm (1⁵⁄₁₆ × 2⁷⁄₈ in.). JPGM 2003.232.

FIGURE 35
Small cast bowl. Late Hellenistic or Roman, 25 B.C.–A.D. 25. H 4 × DIAM 7.5 cm
(1⁹⁄₁₆ × 2¹⁵⁄₁₆ in.). JPGM 2003.237.

FIGURE 36
Cast amber bowl with white ridges. Roman, first century A.D. DIAM 10.3 cm
(4¹⁄₁₆ in.). JPGM 2003.475.

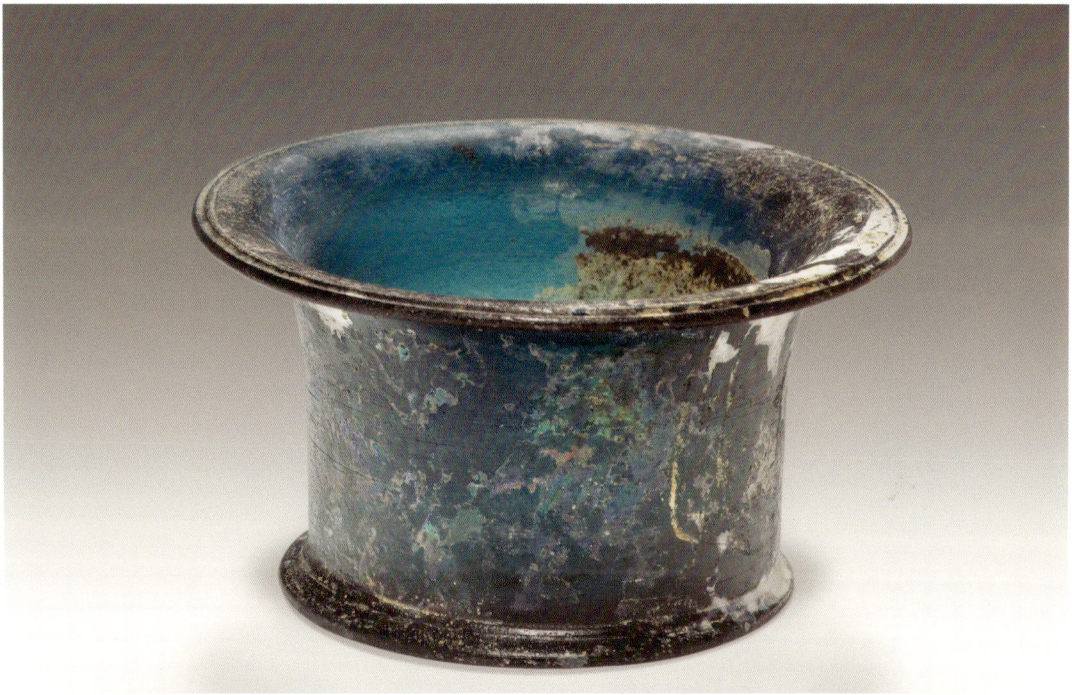

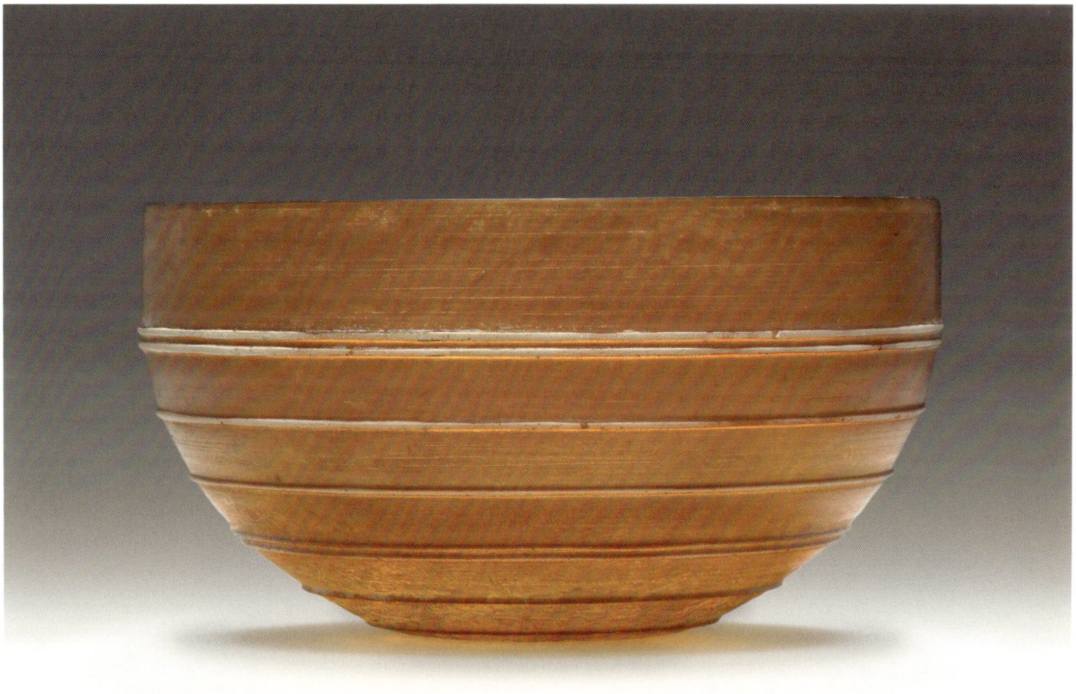

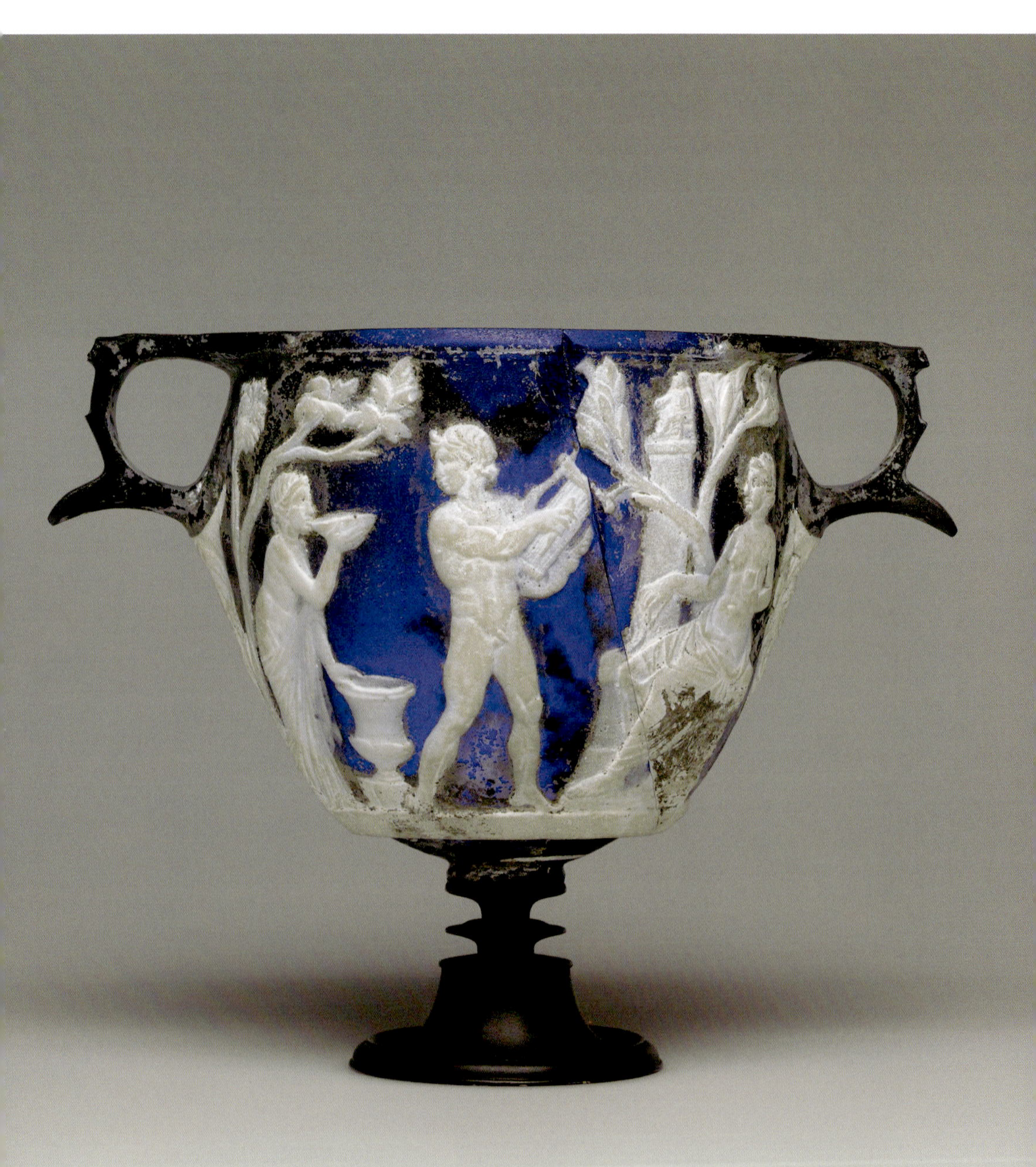

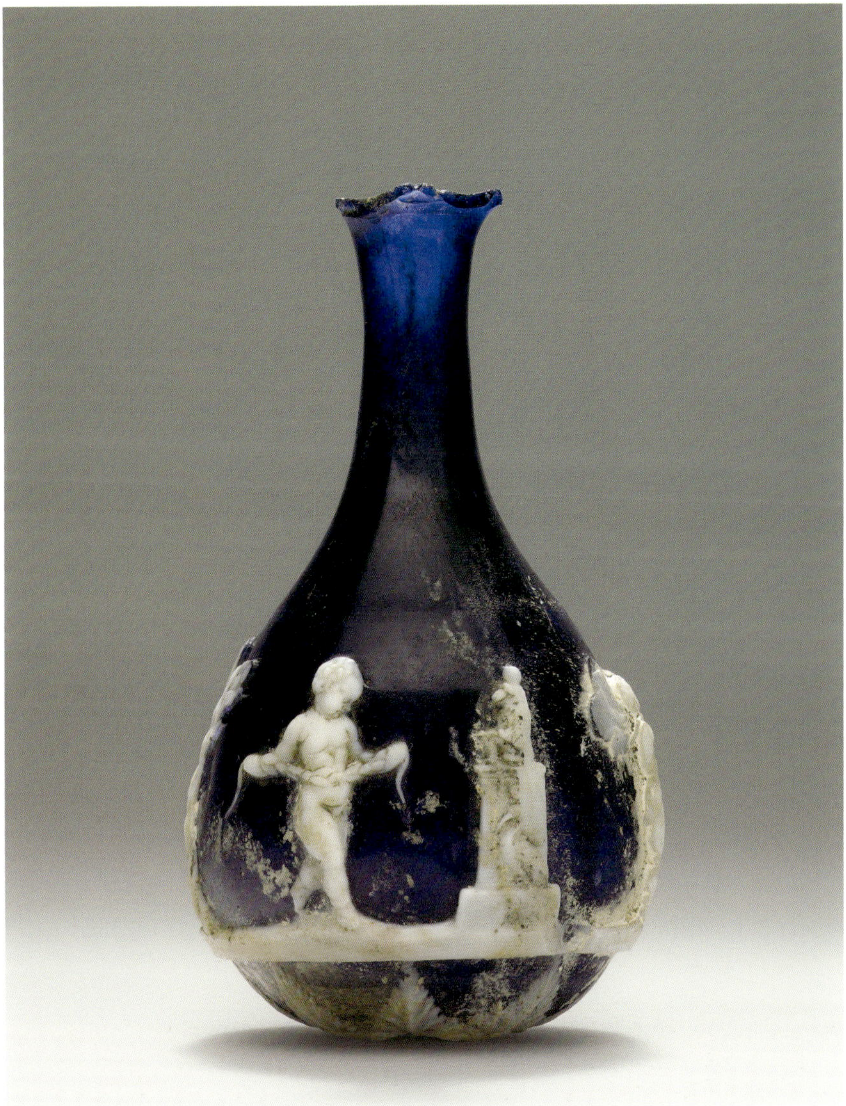

FIGURE 37
Cameo two-handled wine cup (skyphos). Roman, 25 B.C.–A.D. 25. Preserved
H 10.5 × W 17.6 × DIAM 10.6 cm (4⅛ × 6¹⁵⁄₁₆ × 4³⁄₁₆ in.). JPGM 84.AF.85.

FIGURE 38
Cameo perfume flask. Roman, 25 B.C.–A.D. 25. H 7.6 × W 4.2 cm (3 × 1⅝ in.).
JPGM 85.AF.84.

DECORATING BLOWN GLASS

As noted in the previous chapter, the development of the glassblowing technique allowed glassmakers to break free of the conventional shapes and designs of both earlier glasswares and of their cousins in metal and clay. Ingenious shapes and decorative techniques emerged that attest to the ingenuity of these artists. One of the most colorful decorative techniques is known as splashware. After inflating a bubble of hot glass, the glassmaker rolled it into chips of differently colored glass, and after marvering and reheating, these color chips would create splashes of color across the surface of the vessel (Figure 39). A variety of splashware is blobbed ware, in which larger pieces of glass were adhered but not marvered, leaving them in high relief on the surface of the vessel.

Glass in the Roman World

Another decorative technique that resembles the striped decoration found on core-formed glass vessels involved adding trails of glass that were wound around a vessel to create a striped pattern in a contrasting color, but unlike their core-formed predecessors, the trails were not marvered into the body of the vessel but left in relief on the surface. The majority of these vessels have the trail applied symmetrically (Figure 40). Others, however, have patterning referred to by modern scholars as "snake thread" decoration, since the trails of glass arranged over the body take sinuous, snakelike forms (Figure 41). The regional differences in the coloration used on snake-thread vessels are discussed in chapter 6.

Vessels were also decorated by using tools such as jacks, pincers, and tongs to pull and push the glass to create interesting shapes and indentations, or by pinching the glass to make indentations and patterns (Figures 42–44). In some instances, whimsical fish and animal shapes were used for containers (Figure 45). It is tempting to think that a fish-shaped container might have been used to hold *garum*, the famous fermented fish sauce of Roman cuisine. One of the most unusual examples of taking the properties of glass to its fullest extent can be seen in a small number of globular bottles. Their shape is simple, and they are ornamented with a trail of glass of the same color around the neck. But what makes them unique is that the glassmaker used a rod to poke into the bubble he was forming to create interior threads of glass that were connected from one side of the vessel to the other (Figure 46). Exactly why these vessels were made is not clear, but the interior threads may have served a useful purpose. The flask may have contained a liquid that needed to be agitated in order to blend—something like oil and vinegar, perhaps—and the interior threads aided in that agitation.

Another new technique included painted decoration. Once a vessel was shaped and annealed, a painter applied pigment to create a decorative scene (Figure 47). There are two methods known from antiquity for painting vessels. The first is similar to the later **enameling** technique, in which pigment was fired onto the surface of the glass. The pigment itself was suspended in an oily medium. Once the pigment had been applied, the vessel was reintroduced into the glass kiln, but at a lower temperature of 500°C–700°C (930°F–1300°F), which allowed the oily medium to burn away and secure the pigment to the surface of the vessel. Other vessels were painted but not reheated. The decorative scenes painted in this manner are often now worn away due to the fragile nature of the pigment on the surface of the glass. Two shapes were favored by glassmakers of painted vessels. Many, like the example in the Getty Museum, are small wine cups with bucolic scenes of goats, birds, and wreaths. Others are taller beakers made of colorless glass. The finest of the latter group have been found at the site of Begram, in modern Afghanistan, and are painted with scenes of both mythology and daily life.

One unique example of painted glass in the collection of the Corning Museum of Glass is known as the Daphne Ewer, as it depicts the scene from Ovid's *Metamorphoses* (1.452–567) in which the beautiful nymph Daphne, in order to escape the amorous pursuit of the god Apollo, asks Zeus to turn her into a tree (Figure 48). In addition to the color used to describe the scene, gold was added for decorative effect.

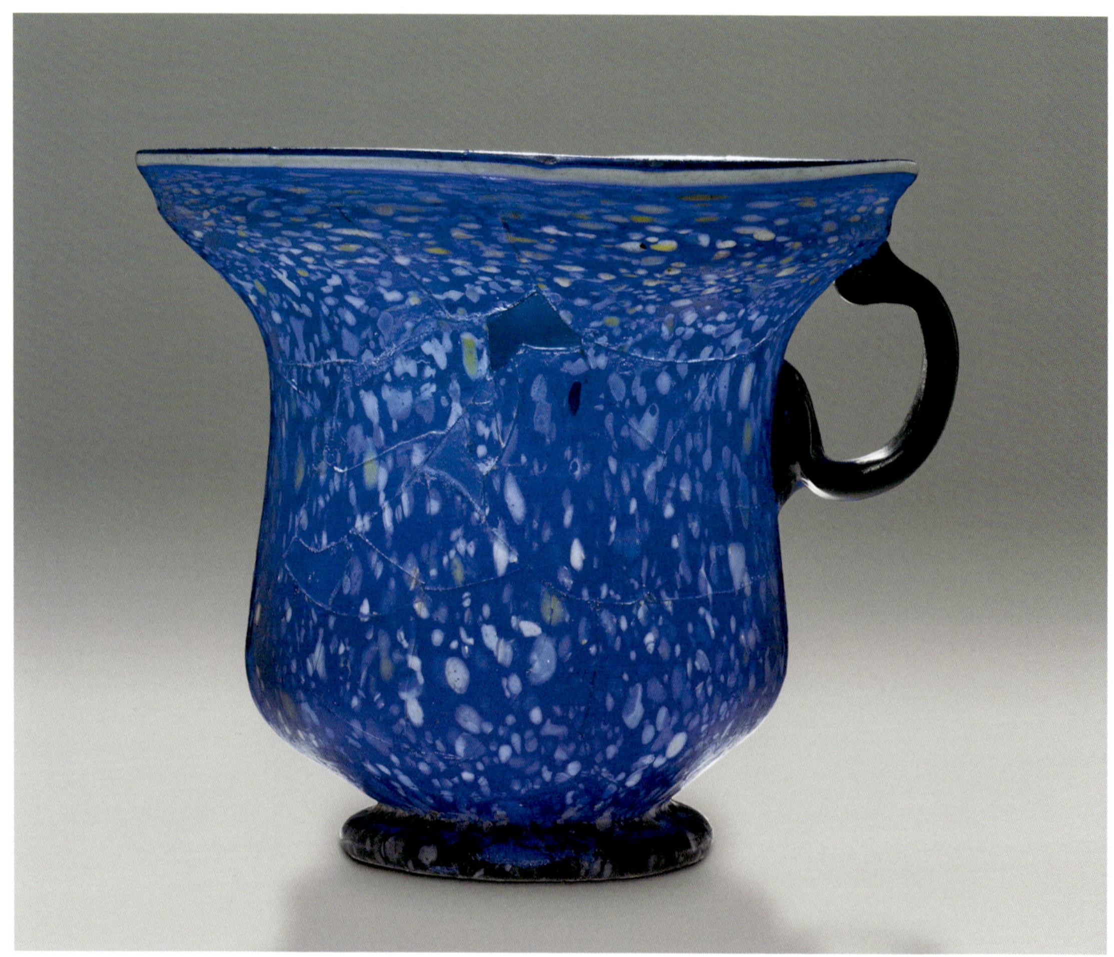

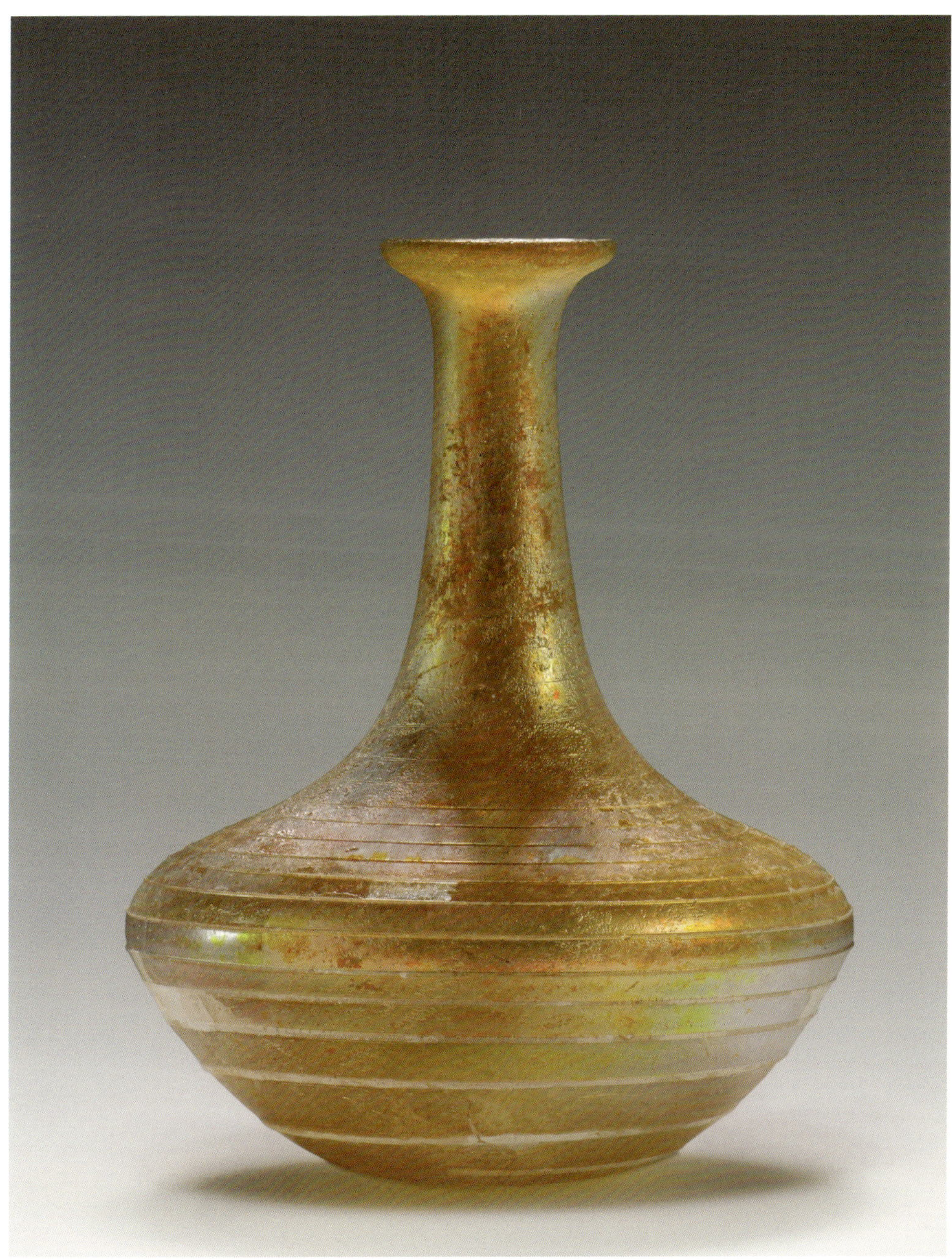

FIGURE 39
Splashware one-handled cup. Roman, first century A.D. H 12.2 × DIAM 14.5 cm
(4¹³⁄₁₆ × 5¹¹⁄₁₆ in.). JPGM 2003.290.

FIGURE 40
Perfume flask with applied trail. Roman, first–second century A.D. H 7.8 × DIAM 6 cm
(3¹⁄₁₆ × 2³⁄₈ in.). JPGM 2003.420.

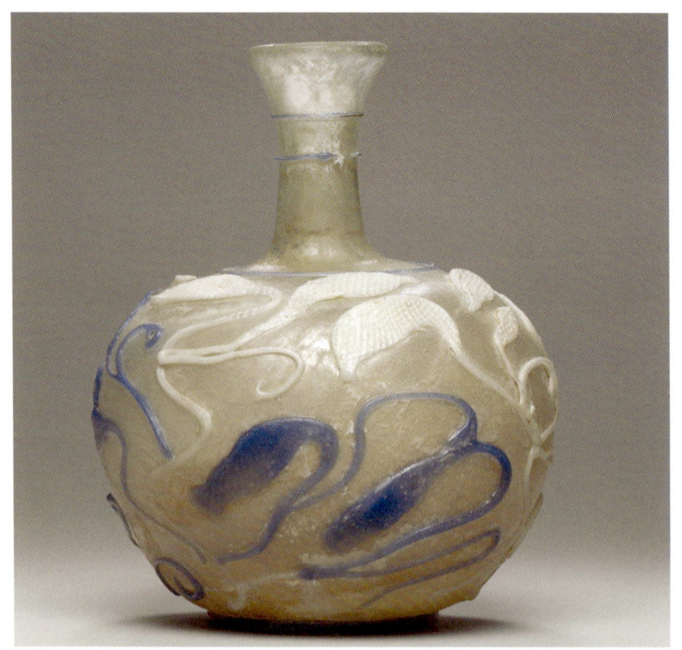

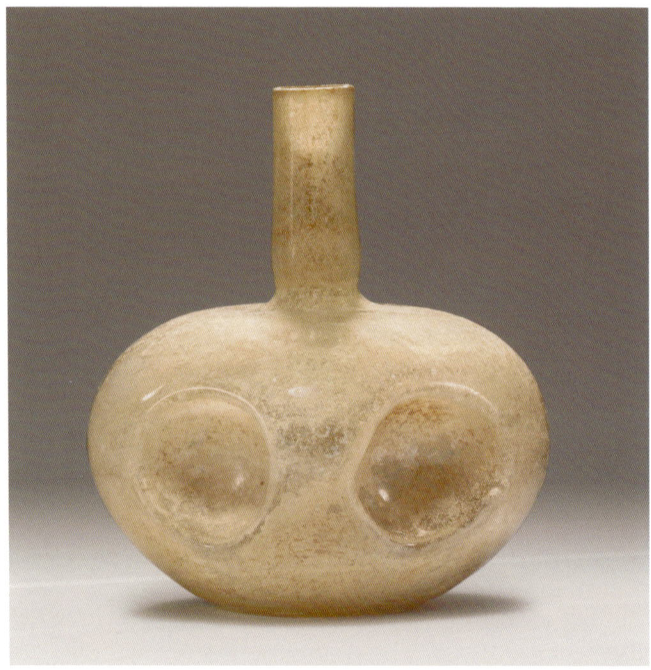

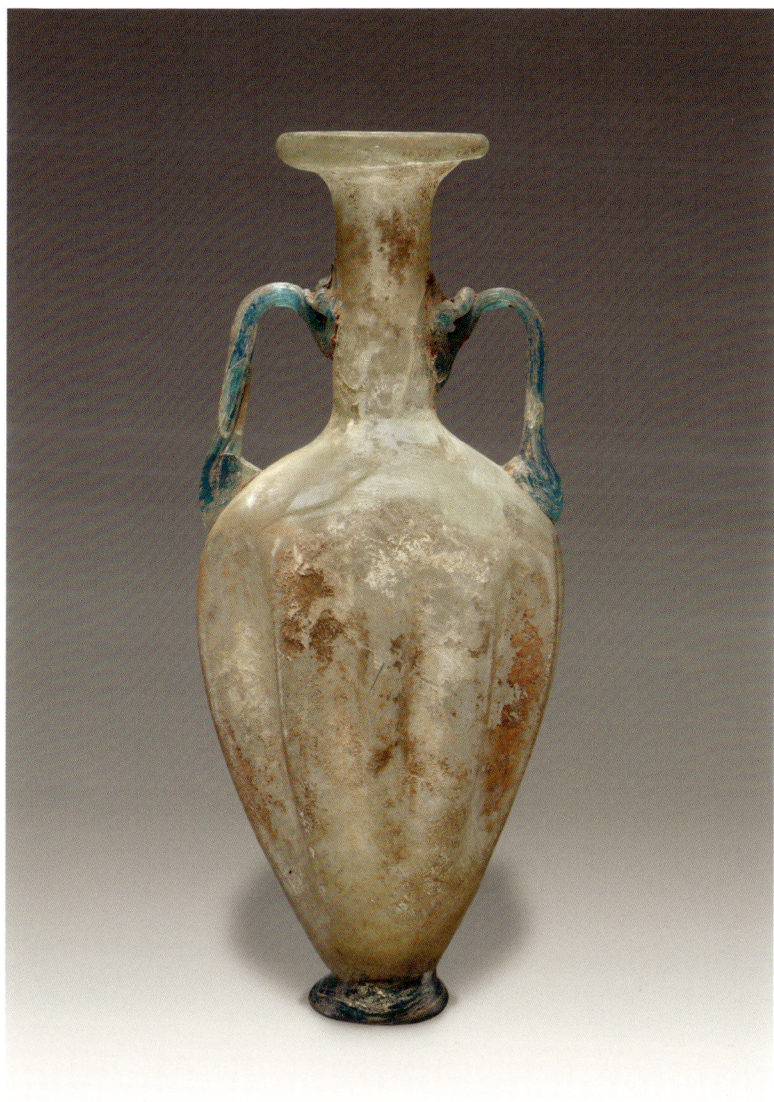

FIGURE 41

Snake-thread flask. Roman, third century A.D. H 14.2 × Mouth DIAM 3.6 cm
(5⁹⁄₁₆ × 1⁷⁄₁₆ in.). JPGM 96.AF.56.

FIGURE 42

Flask with circular indentations. Roman, third century A.D. H 10.5 × DIAM 10 cm
(4¹⁄₈ × 3¹⁵⁄₁₆ in.). JPGM 2003.408.

FIGURE 43

Small storage jar (amphoriskos) with vertical indentations. Roman, fourth century A.D.
H 19.3 × DIAM 8.5 cm (7⁵⁄₈ × 3³⁄₈ in.). JPGM 2003.414.

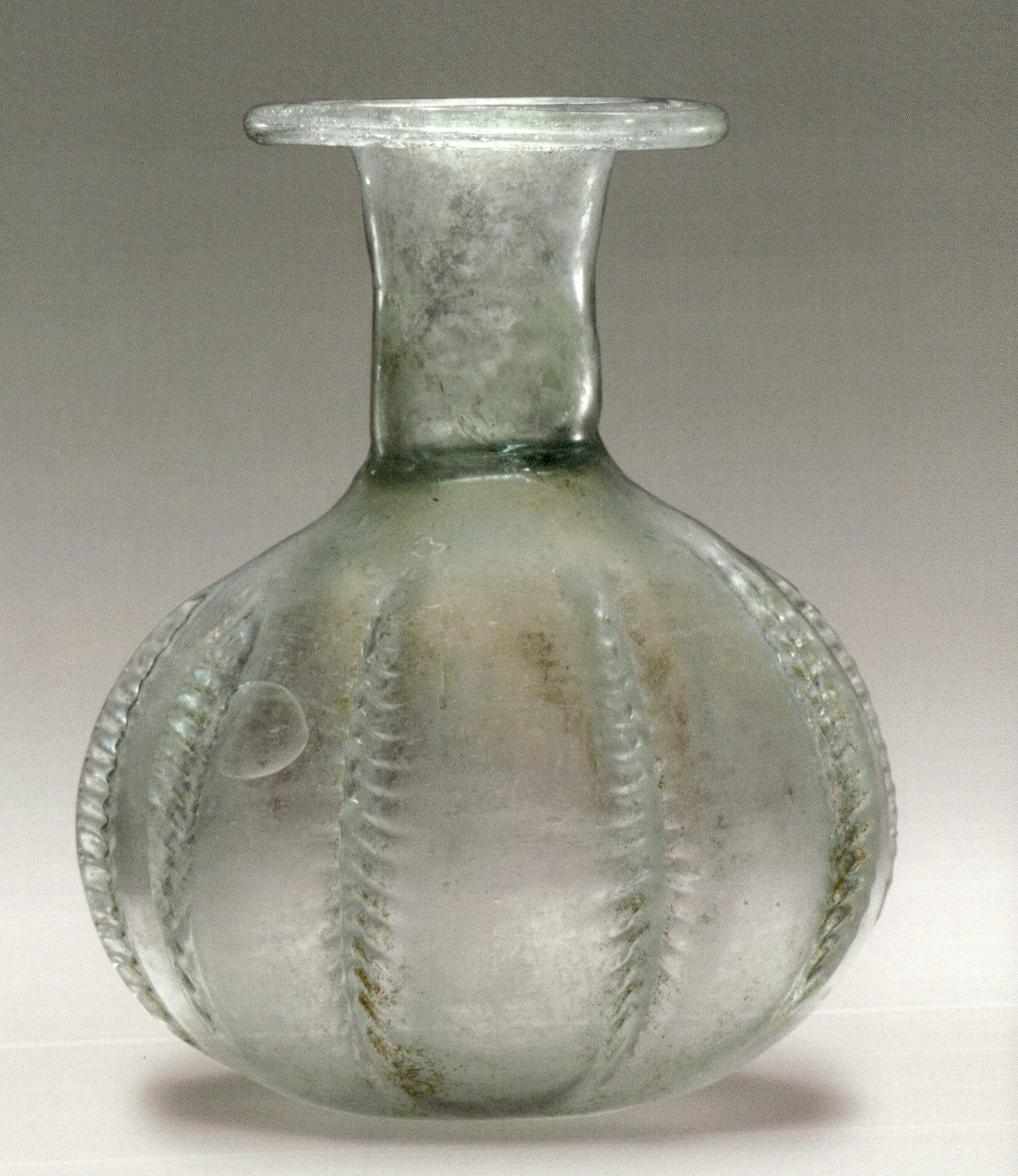

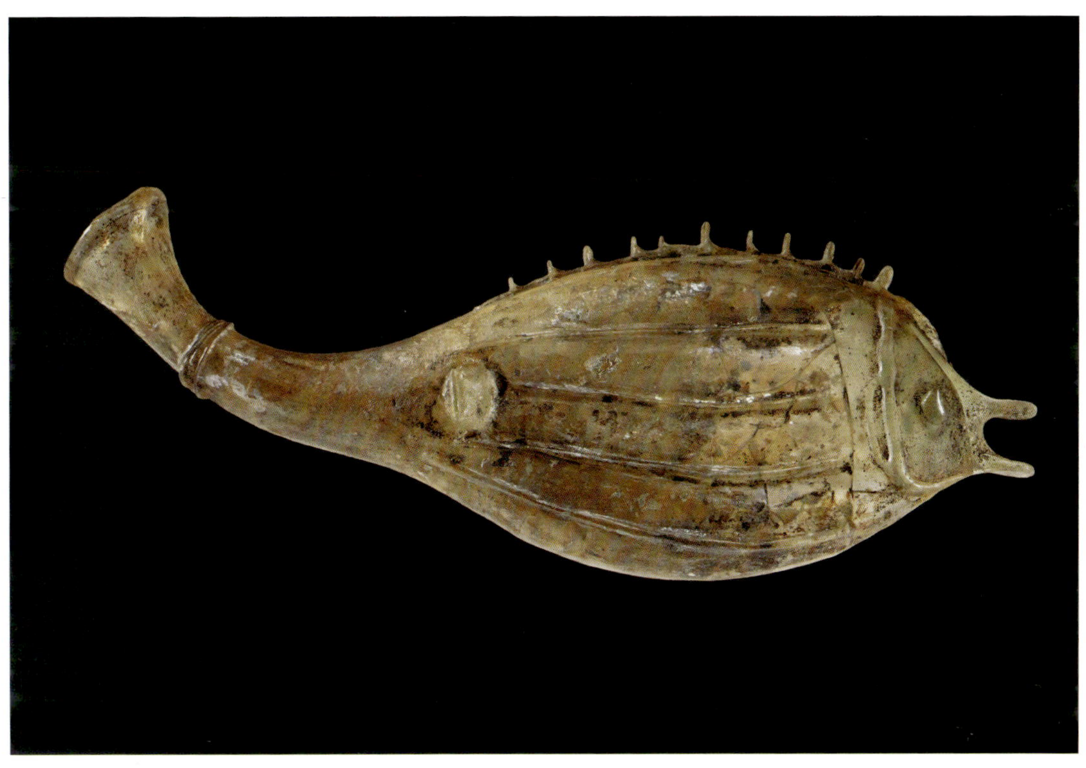

FIGURE 44

Sprinkler flask with pinched vertical ribs. Roman, third–fourth century A.D.
H 9.5 × DIAM 7.5 cm (3¾ × 2¹⁵⁄₁₆ in.). JPGM 2004.42.

FIGURE 45

Flask in the form of a fish. Roman, third century A.D. H 8 × L 21 cm (3⅛ × 8¼ in.).
JPGM 2003.439.

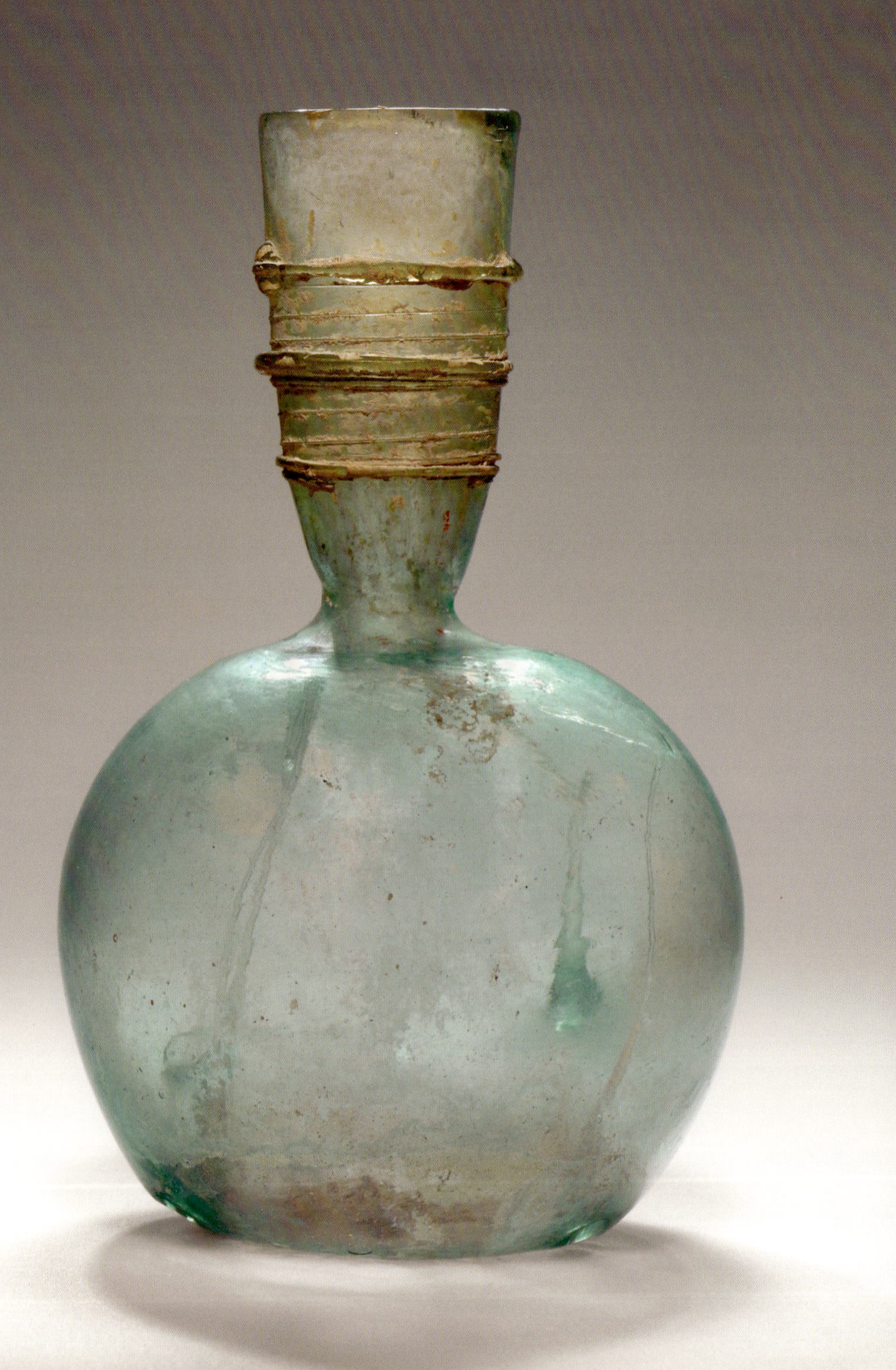

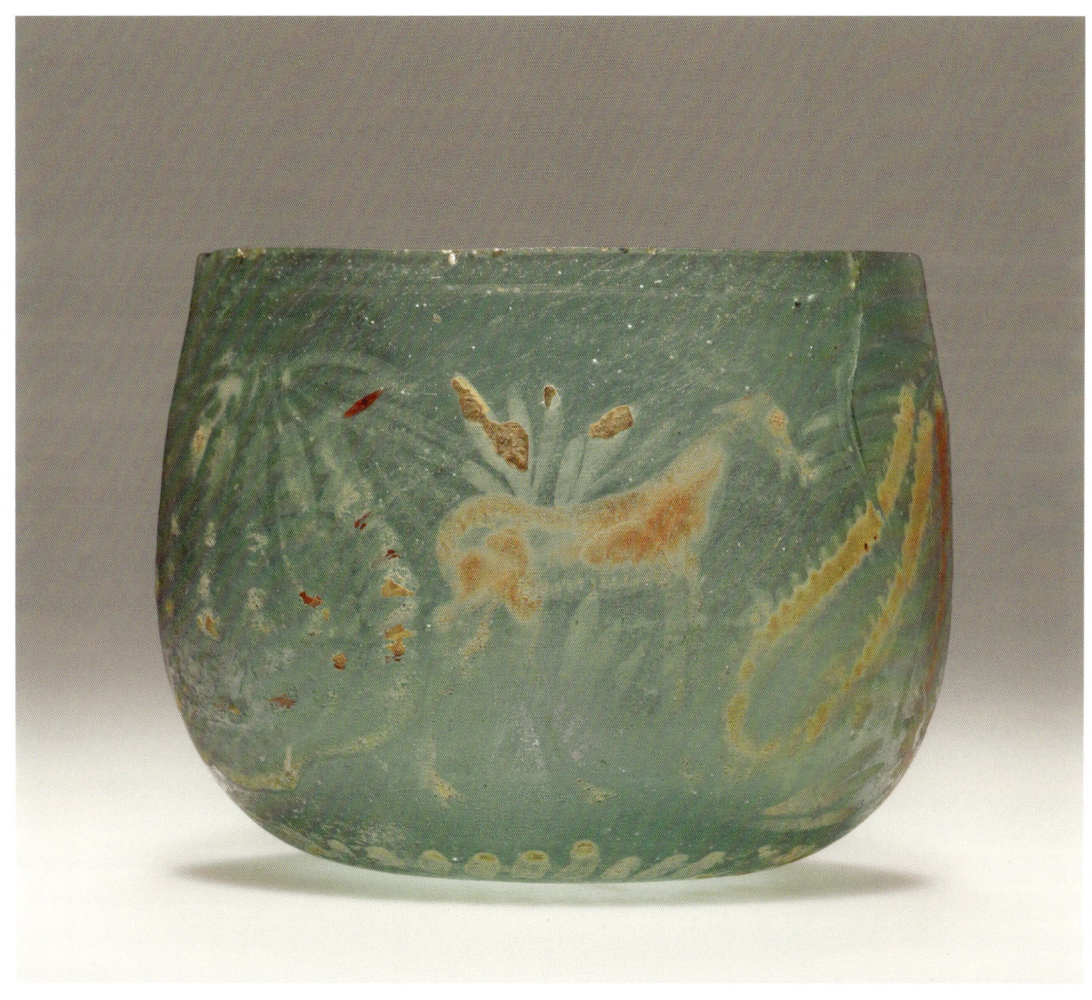

FIGURE 46
Flask with interior threads. Roman, fourth century A.D. H 13.6 cm (5³⁄₈ in.). JPGM 71.AF.85.

FIGURE 47
Wine cup with enameled bucolic scene. Roman, first century A.D. H 6.5 × DIAM 8 cm
(2⁹⁄₁₆ × 3¹⁄₈ in.). JPGM 2003.294.

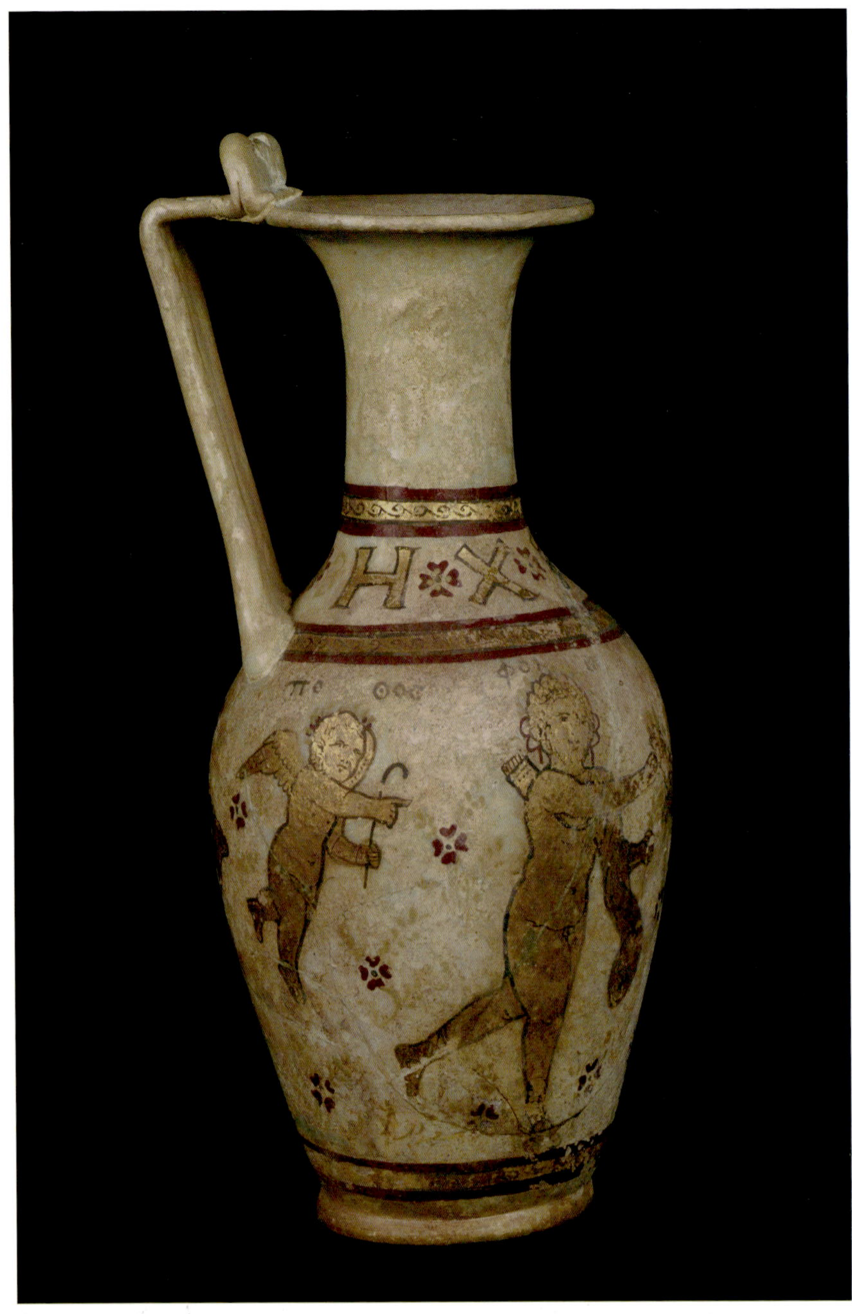

FIGURE 48

The Daphne Ewer. Roman, late second–third century A.D. H 22.2 × DIAM 9.5 cm (8¾ × 3¾ in.).
CMOG 55.1.86. Photo: CMOG.

MOLD-BLOWN GLASS

Although glassmakers began pushing the limits of their medium in the first century, they also continued to make vessels that maintained a close connection with ceramic and metal forms. Beginning in the late first century B.C. and early first century A.D., glass vessels were manufactured by inflating the bubble at the end of the blowpipe into a mold carved with a relief pattern (Figure 49). Molds had been used to make cast wares for quite a long time, thus it was a natural extension to use them now for inflated wares. What was different, however, was the complexity of the shapes and designs of the molded glasswares and the molds used to create it.

Many of the molds used for the production of ceramic pieces were designed with wide openings. This enabled the ceramist first to insert the wet clay and then to remove it once it had dried and shrunk away from the walls of the mold. Similar molds were used for glass production, and those constructed as a single piece are called **dip molds**. They are used to make open bowls and beakers, and their decorative patterns are shallow and do not hinder the glass as it is pulled out of the mold. Dip molds are also used to impart a pattern that could then be exploited through further inflation and manipulation. But other molds were much more complex and had high relief patterns. They were made of two or more parts.

In using molds, a glassmaker may have had an assistant who would close the mold around the glass bubble and hold it in place while the glass was inflated. The glassmaker had to blow hard to force the hot glass against the walls of the mold to allow it to take on the relief decoration. Once the glass bubble was fully inflated, the mold had to be removed so that the glass did not adhere to the mold as it cooled. This was a delicate process because the molded vessel could easily be distorted in the process. Many glassmakers today are able to inflate glass into simple two-part molds without the need for an assistant, provided the mold is designed to come apart easily.

The very few fragments of glass molds that have survived from antiquity are from molds constructed in two halves. The materials used to make these molds are clay and stone, although glassmakers in later periods (and today) used plaster as well, and it is likely that ancient glassmakers did too. A bronze six-sided cup found in Samaria that dates to the fifth century A.D. may have been used as a mold for a series of small pitchers, because glass vessels have been found with a pattern identical to that of the metal cup. The idea of ancient glassmakers using metal molds is reinforced by the existence of a unique cup dating to the first century A.D. and now in the collection of the British Museum (inv. no. GR 1870.9-1.2). It is composed of cobalt blue glass that was inflated into a silver container designed with a series of ovals cut out of the metal. As the blue glass expanded, it bulged out through the oval holes to create a beautiful pattern as well as a pleasing surface texture that provided the person holding the cup a better grip. It must have been a delicate procedure to make the vessel, as the heat of the glass and its silver container must have been carefully controlled to avoid thermal shock between the two materials.

Scholars are able to determine the number of mold elements used to make a vessel by examining the seam marks left in the glass. As it was inflated into the mold, the hot glass was forced into the mold seams, leaving behind a telltale line. The mold makers for the glass quickly realized this and began to pattern their molds so that the seams became part of the design—the stem of a plant, or an architectural column, for example—thus hiding the mold seam in the vessel's decoration. A small wine cup in the collection of the Getty Museum exemplifies this technique (Figures 50a, b). The mold pattern contains an inscription in Greek divided between the two sides, with the word KATAXAIPE ("Greetings to all!") on one side, and KAI ΕΥΦΡΑΙΝΟΥ ("Be of good cheer!") on the other. The two phrases are separated by vertical palm fronds that help to conceal the seams of the halves of the mold that formed the upper body of the cup. This cup was made using a mold constructed of three parts: the two side elements just described form the vertical walls of the cup, while the third is the bottom of the cup and incorporates the chevron pattern that encircles the bottom of the vessel. It was the assistant's job to keep these mold parts registered properly and held tightly together while the glass was inflated. But imagine doing this with a mold constructed of eight side panels and a base element! It is unfortunate that none of the more complex molds that were used to construct vessels of great size, diversity, and complexity have survived, as they could have shed much light on the techniques and manufacture of mold-blown glass. In fact, they are so complicated that modern glassmakers have not attempted to replicate them. By the end of the first century A.D., the complicated mold designs that incorporated multiple panels of decoration had given way to simple mold designs of only two parts.

The use of molds to manufacture glass vessels meant that numerous replicas could now be made, often using the same color, but sometimes using glass of differing colors, which meant that even within a known shape or pattern some variety existed. Two small perfume flasks in the collection of the Getty Museum, one of opaque white glass and the other transparent blue, illustrate this point well. Both were made using a hexagonal mold whose design consisted of a series of different vase shapes, hence they are called "vessel" flasks by modern scholars (Figure 51). It is most likely that when the glassmaker was blowing "vessel" flasks, he used the same color of glass over and over until the batch had been used up. Only then would another color start to be produced. It all depended upon what color of molten glass was sitting in the crucibles inside the glass kiln. The time needed to manufacture such a small bottle was minimal, a matter of only a few minutes—gather the glass, blow a small bubble, place the mold around it, inflate the glass, remove the mold, separate the bottle from the blowpipe, and finish off the rim. All the bottles could be annealed together, thus when they were ready to be sold, a variety of colors was available, and customers could select their favorite. Alternatively, it may be that since these small flasks were used to hold scented oil, the perfumer made the decision that a certain scent would be sold in one color, and another in a different color, enabling customers to make sure they were purchasing the scent they wanted.

The mold patterns were designed with great variety and humor. They also contain the only names preserved from antiquity of those artisans associated with glassmaking. Among the most prolific was a man called Ennion. His workshop produced a variety of wares, from tall pitchers to small cups and cosmetic containers that have come from archaeological contexts that date from the A.D. 30s and 40s. Many have been found in northern Italy, and it is possible that Ennion started his career in glassmaking somewhere in the eastern Roman Empire and then relocated to Italy in search of new markets (Figure 52). When looking at his wares, it is hard to visualize how the complex molds were assembled and the vessels made. What is characteristic of his mold designs is the manner in which his "signature" is recorded. Within a small *tabula ansata* is a phrase in Greek that reads ΕΝΝΙΩΝ ΕΠΟΙΗCΕΝ ("Ennion made [it]"). What remains unclear to scholars today, however, is whether Ennion was the maker of the mold, the glassmaker, or perhaps the proprietor of the glass workshop? Scholars have tended to assign him the role of proprietor, since similar "signatures" exist on Roman ceramics where the identity and role of the person named is clear. The fact that the inscriptions on his vessels are written in Greek indicates his origins in the Greek-speaking regions of the eastern Roman Empire.

Ennion's vessels are characterized by their use of decorative patterning that is carried over from Hellenistic metal and ceramic work. The cup pictured in Figure 52, for example, is decorated on its body with vertical flutes or tongues, a zigzag pattern on the base, and foliate tendril friezes on either side of the inscriptions. Almost all the products of Ennion's workshop have panels containing his name. His competitors used the same method to advertise their wares, thus we know the names of Neikais, Jason, Meges, and Aristeas (Figure 53). Other proprietors and firms incorporated their names into the base molds of round and square bottles used to transport liquid goods around the empire, designs that often included the figure of Mercury, the Roman god of commerce and trade, a fitting symbol for such commodities.

In designing their patterns, mold makers frequently resorted to reusing patterns found on ceramic and metal vessels or copied shapes that had been in use for centuries, such as animal and human heads (Figures 54, 56). But they also turned to nature for their designs, creating vessels that take the forms of fruits, vegetables, and bunches of grapes (Figure 55). In some instances, it is thought that the mold itself was made by pressing the clay or plaster against the fruit.

Still other mold designs were unique to glassware, which showed that the mold makers were also thinking of glass as a unique medium that could hold a pattern that clay could not, and they were not completely tied to the standard designs used for ceramics and metalware. In some cases, designs were developed for one shape, such as a cylindrical cup in the collection of the Getty Museum with a repeating pattern of pomegranate fruits alternating with stalks of grain, that then were repeated again on a vessel of a different shape (Figure 57). Presumably these vessels with different shapes but the same pattern were made in the same workshop in order to provide customers both with different options as well as with an opportunity to buy sets of glassware. We need only to think of the way we purchase drinking vessels, in four-, six-, and eight-ounce containers made of the same design to understand why such vessels of similar pattern but different size and shape were made in antiquity.

Other examples of designs created exclusively for glass can be seen in the many tall beakers used for drinking wine (Figures 58–61). One variety that contains knobby protrusions has been called lotus bud beakers, as the knobs resemble the flower buds. Lotus bud beakers were extremely popular in antiquity and examples in various sizes, both short and tall, have been found throughout the empire. Another variety, called mythological beakers, is decorated on the exterior with figures of four gods or mythical personifications that stand like statues, separated by columns. Like the lotus bud beakers, mythological beakers have been found throughout the breadth of the Roman Empire, from Dura Europas in Syria to Nijmegen in Belgium. It is likely that some of these wares were being transported by the armies who marched their way across the provinces.

One technique that came back into fashion in the late first century A.D. and reached its ultimate heights in the fourth century is that of cutting and engraving. While cast vessels from centuries earlier had been decorated with cut designs, nothing matched what the Romans did. During the Flavian period (A.D. 69–96), a method of facet cutting on the exterior of vessels was devised (Figure 62). This decorative patterning coincided with the introduction of deliberately colorless glass and was probably meant to imitate rock crystal (see Figure 13). A similar approach to colorless glass had been taken centuries earlier in the Achaemenid world. The technique of facet cutting used on drinking vessels is the same today as it was in antiquity, and one need only think of the cut crystal that became fashionable in the Victorian period to see the similarities. Other decorative patterns were also incised on vessels, some quite elaborately, with combinations of geometric patterns or figures, florals, and inscriptions (Figures 63, 64). These wares are different from the faceted drinking cups, however—the cutting is very shallow and sketchy in comparison.

But the ultimate products of the cutting technique were the cage cups, thick blanks of glass that were cut and undercut to create a cagelike network of glass on their exteriors (Figure 65). These impressive vessels were made by first casting thick-walled blanks that were then handed over to the cutting specialists. The glass cutter carved a netlike pattern by cutting deeply into the blank to create a series of interlocking rings that stand proud of the vessel's inner wall but are connected to it by thin struts, or bridges, of glass. It is an extremely time-consuming and challenging process, and one in which any little mistake would result in disaster. Some of the blanks used for carving cage cups were multicolored. The design of the cut decoration on occasion included inscriptions around the rims of the vessels that were a different color both from the interior and from the glass below it in the design (Figure 66). A stunning example found in a Roman burial in ancient Colonia Claudia Ara Agripinnensium, the present-day city of Cologne, Germany, has a Greek inscription encircling the rim: ΠΙΕ ΖΗCΑΙC ΚΑΛWC ΑΕΙ ("Drink! May you live for many years!"). A similar example found in Novara, in northwestern Italy, has the same inscription, but in Latin. Other cage cups were monochromatic, but no less spectacular. While the majority are designed with a conical shape, two have survived that are hemispherical and may have been used as suspended lamps. An example at the Corning Museum of Glass still has a bronze circular band around its rim and chains for suspension (see Figure 65).

The most elaborate of the cage vessels is the so-called Lycurgus Cup in the British Museum (inv. no. M & ME 1958,12-2,1). Rather than having a pattern of interwoven concentric rings, it has a figural scene showing the myth of King Lycurgus, who was entangled in grapevines after angering Dionysos, the Greek god of wine. In addition to its rare figural design, the Lycurgus Cup was made using an unusual type of glass known as dichroic, in which minute traces of gold and silver are included in the glass batch. Their reflective qualities contrive to change the glass color with the direction of the light source. The Lycurgus Cup, thus, was made of an opaque glass that is green in reflected light but red in transmitted light. One can only wonder if dichroic glass was an accidental discovery or a deliberate experiment in finding new ways to color glass. In either case, it provided another opportunity for glassmakers to exploit the unique qualities of the medium and achieve spectacular results. Dichroic vessels, like cage cups, are very rare.

Very few cage cups have survived from antiquity, which suggests that like the cameo glass of the Augustan age, it was reserved for the highest levels of society. The majority of the surviving examples have been found in central Europe, but fragments of others have been found around the Mediterranean, including a fragment in Carthage.

By the mid-first century A.D., the capabilities of glassmakers had expanded exponentially compared to the previous centuries of production. A revolutionary technique had been developed that suddenly allowed the imagination of glassmakers to expand across the empire as new shapes, fresh decorative methods, and mass quantities of glass spread far and wide. Cast, mosaic, and core-formed vessels had been replaced by inflated wares decorated in a variety of stunning techniques, and because of this plethora of wares, taste and fashion dictated glass production.

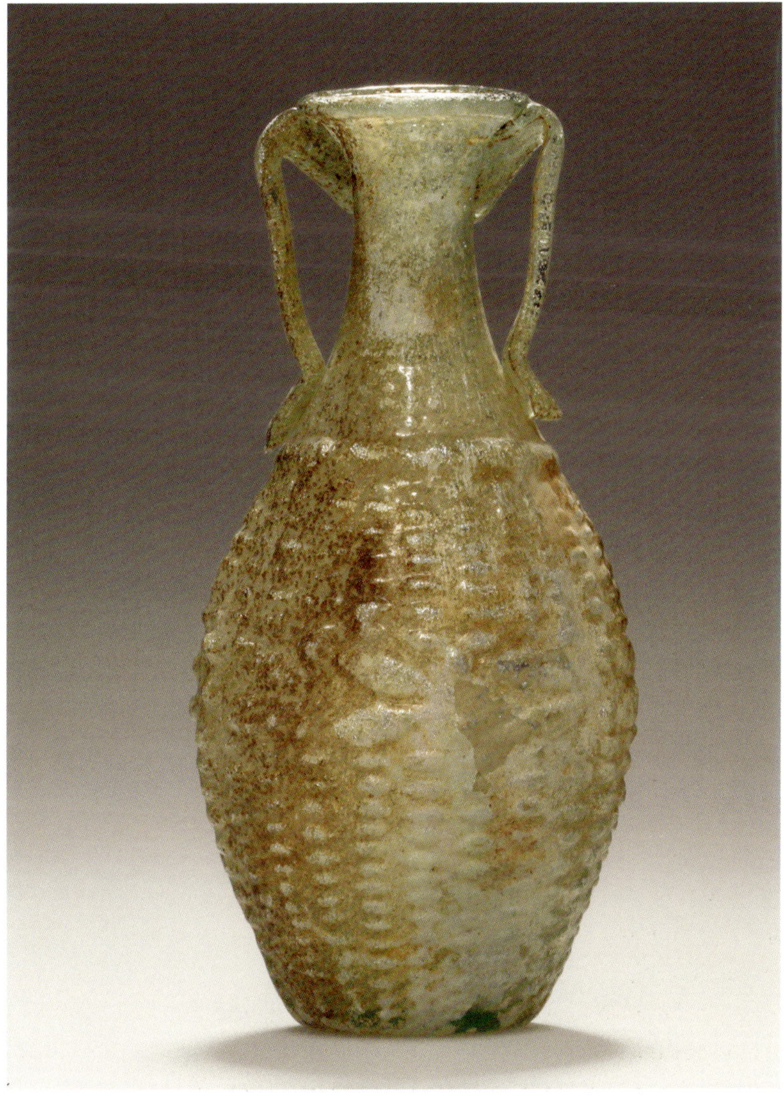

FIGURE 49

Mold-blown flask shaped as a woven basket, attributed to the Workshop of the Floating Handles. Roman, first century A.D. H 9.2 cm (3⅝ in.). JPGM 2003.309.

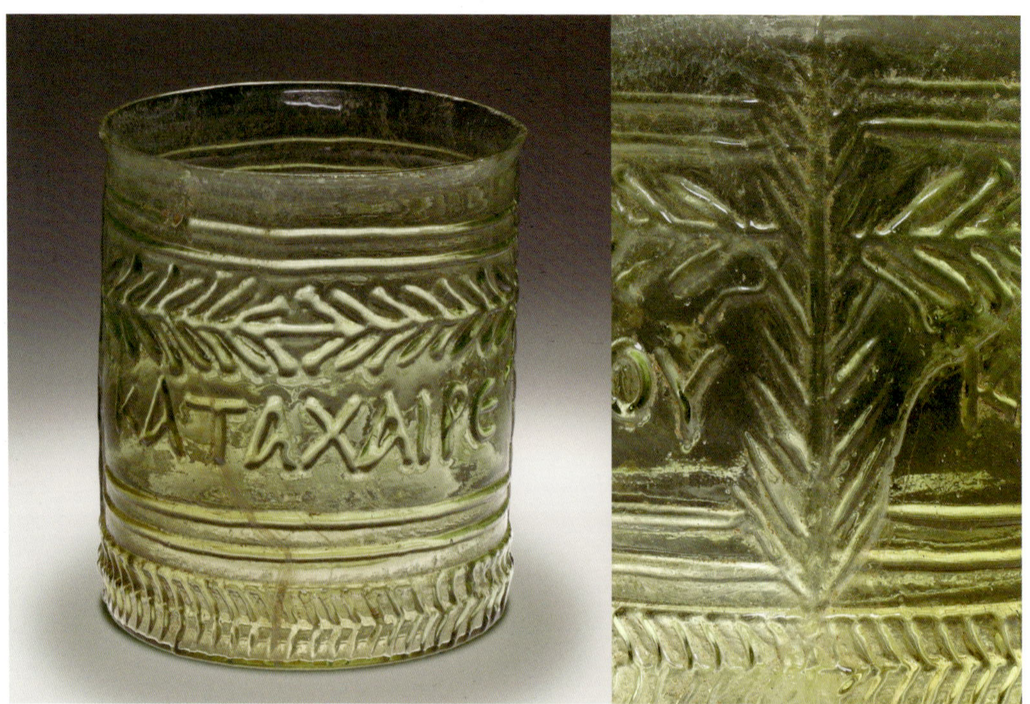

50A.   50B.

FIGURE 50A, B

A. Mold-blown wine cup with an inscription. **B**. Detail of the concealment of the mold's seam mark within the pattern of the design. In this instance, the mold seam becomes the central spine of the palm frond. Roman, first century A.D. H 7.8 × DIAM 7.3 cm (3 1/16 × 2 7/8 in.). JPGM 2004.35.

FIGURE 51

Two mold-blown "vessel" flasks. Roman, first century A.D. (*left*) H 8.4 cm (3 5/16 in.). JPGM 2003.299, (*right*) H 7 cm (2 3/4 in.). JPGM 2003.297.

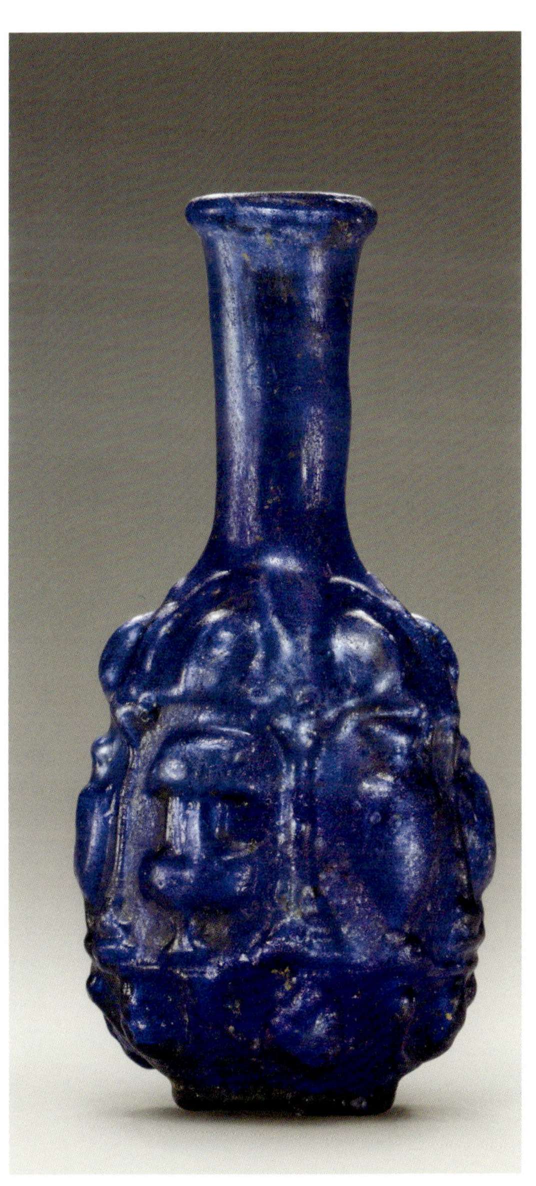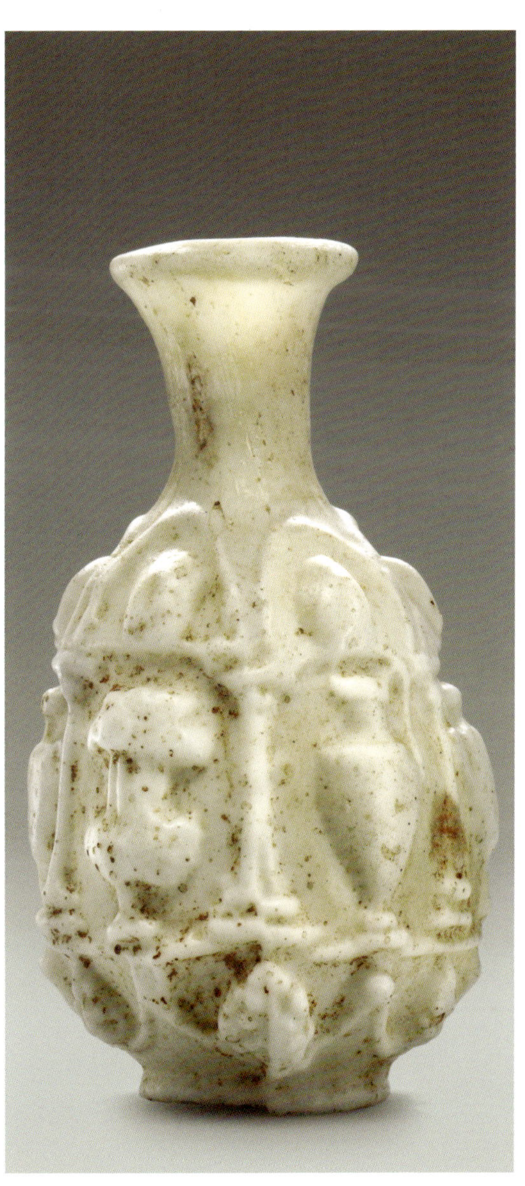

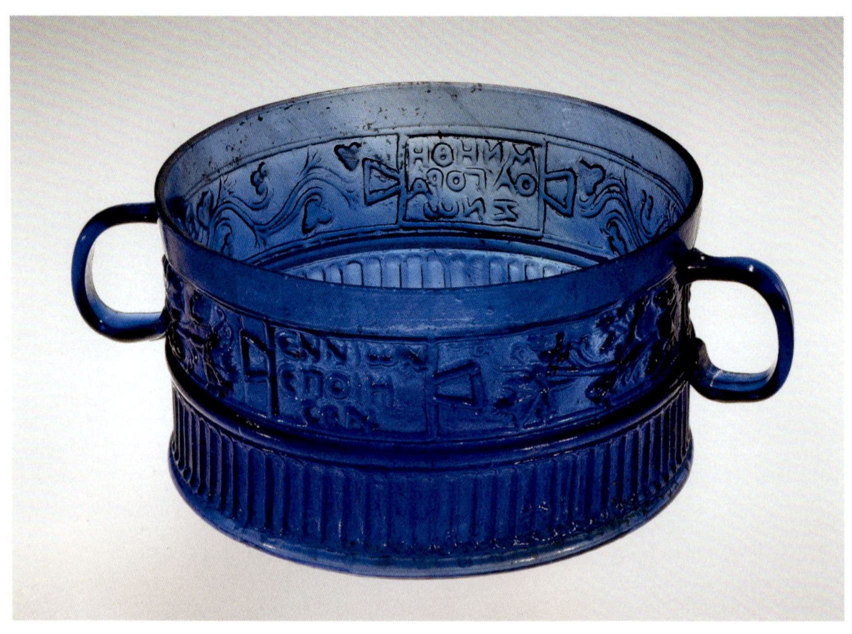

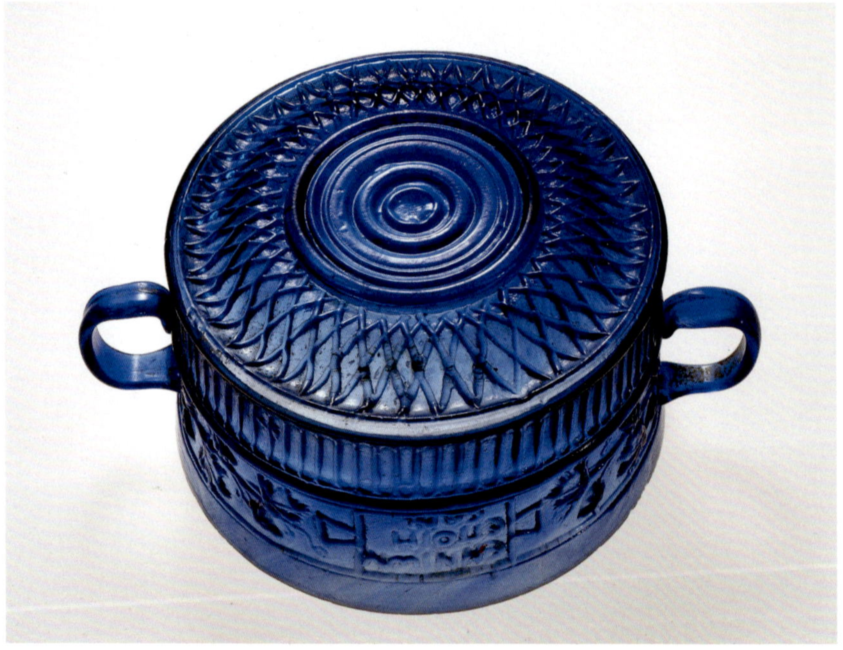

FIGURE 52

Mold-blown two-handled wine cup with an inscription reading, "Ennion made [it]." Roman, mid-first century A.D. H 6 × DIAM 9.5 cm (2³⁄₈ × 3³⁄₄ in.). CMOG 66.1.36. Photo: CMOG.

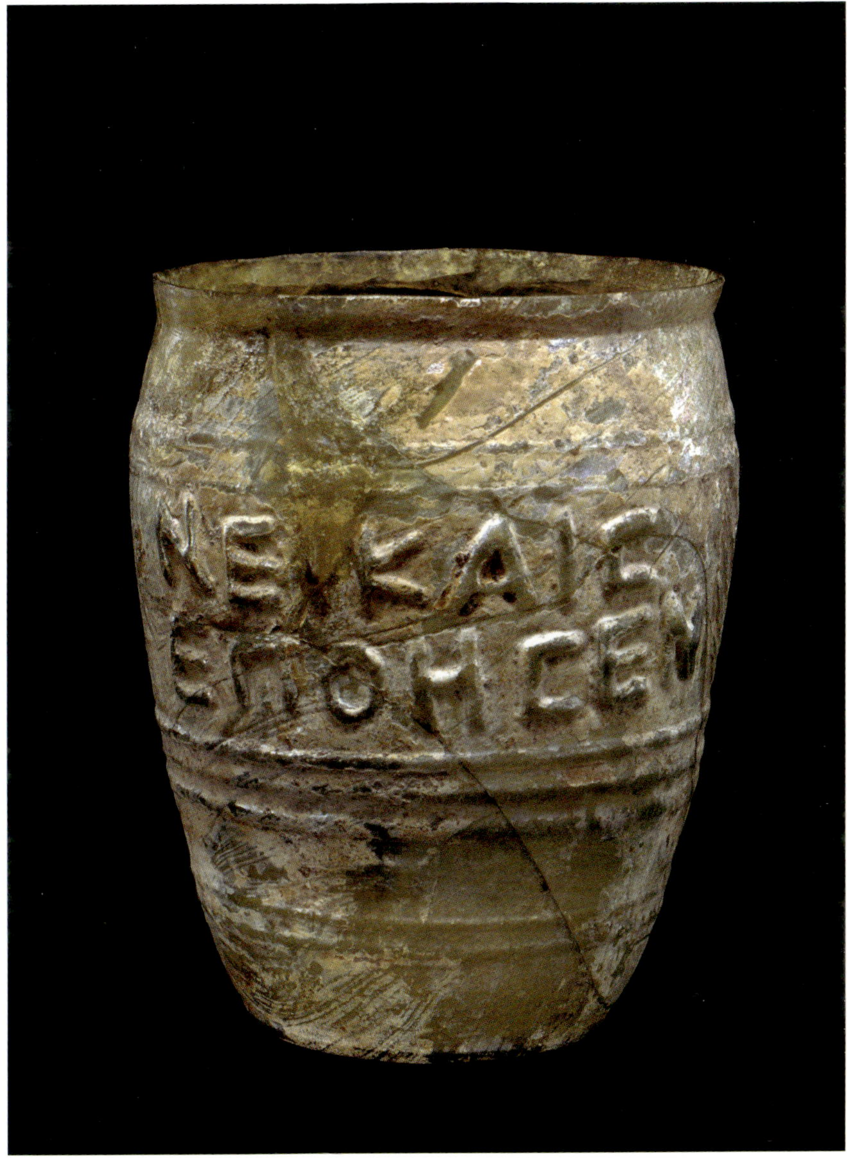

FIGURE 53
Mold-blown wine cup with an inscription reading, "Neikais made it." Roman, mid-first century A.D.
H 8.7 × Rim DIAM (as restored) 6.5 cm (3³⁄₈ × 2¹⁄₂ in.). CMOG 70.1.8. Photo: CMOG.

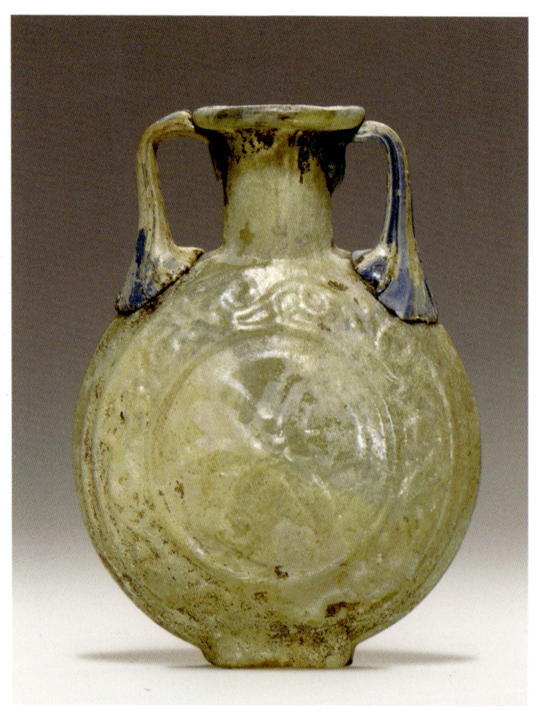

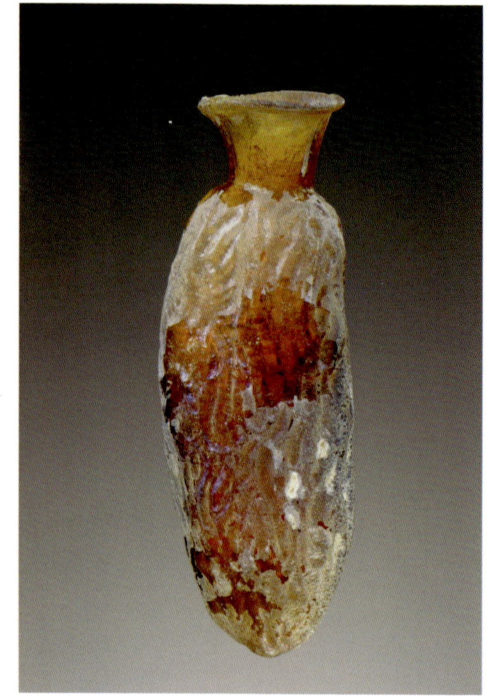

FIGURE 54
Mold-blown flask with rosettes. Roman, first century A.D. H 7.6 cm (3 in.).
JPGM 2003.311.

FIGURE 55
Mold-blown date flask. Roman, first–second century A.D. H 7 cm (2¾ in.).
JPGM 2003.334.

FIGURE 56
Mold-blown head flask. Roman, second century A.D. H 19.4 cm (7⅝ in.).
JPGM 2003.326.

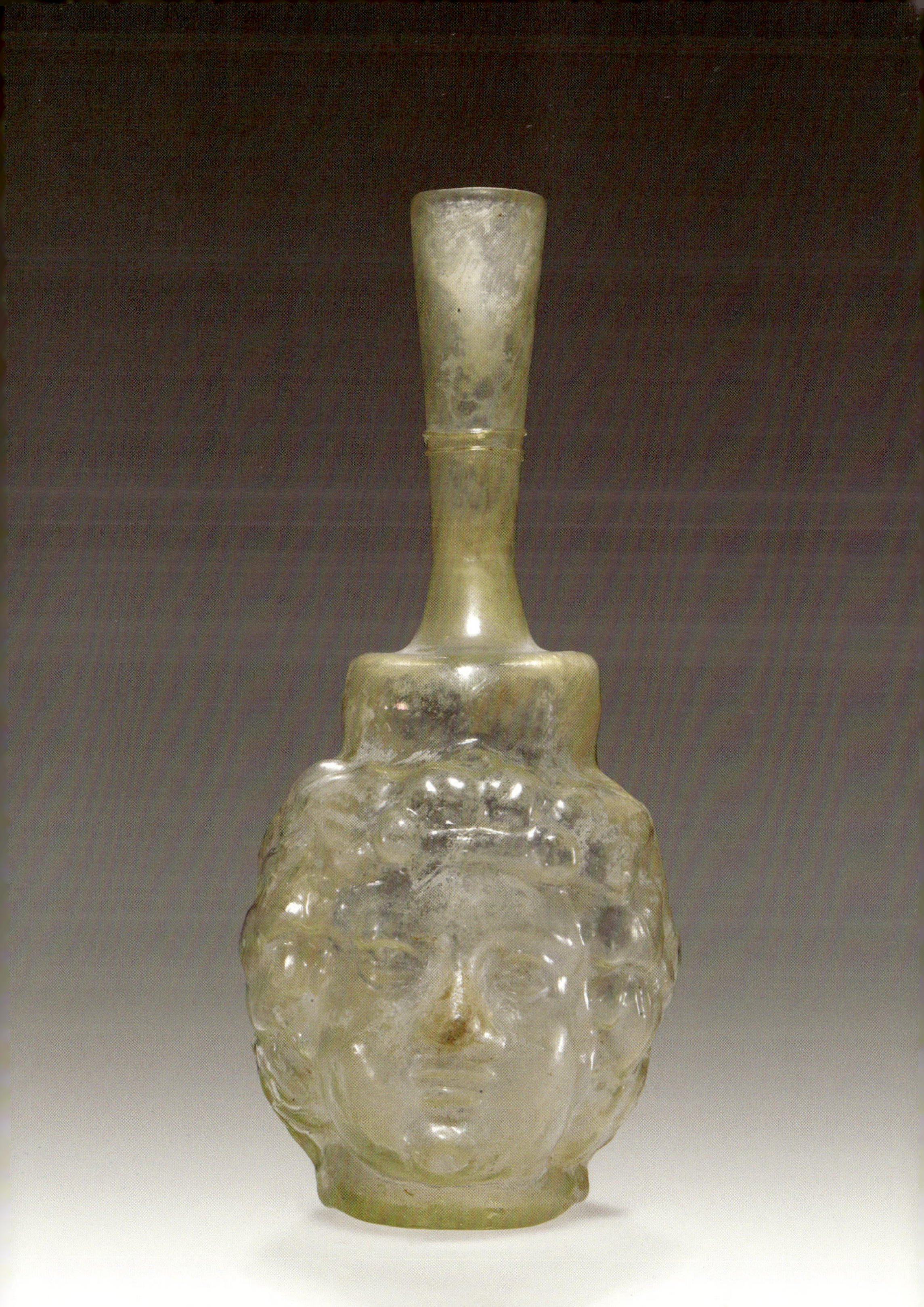

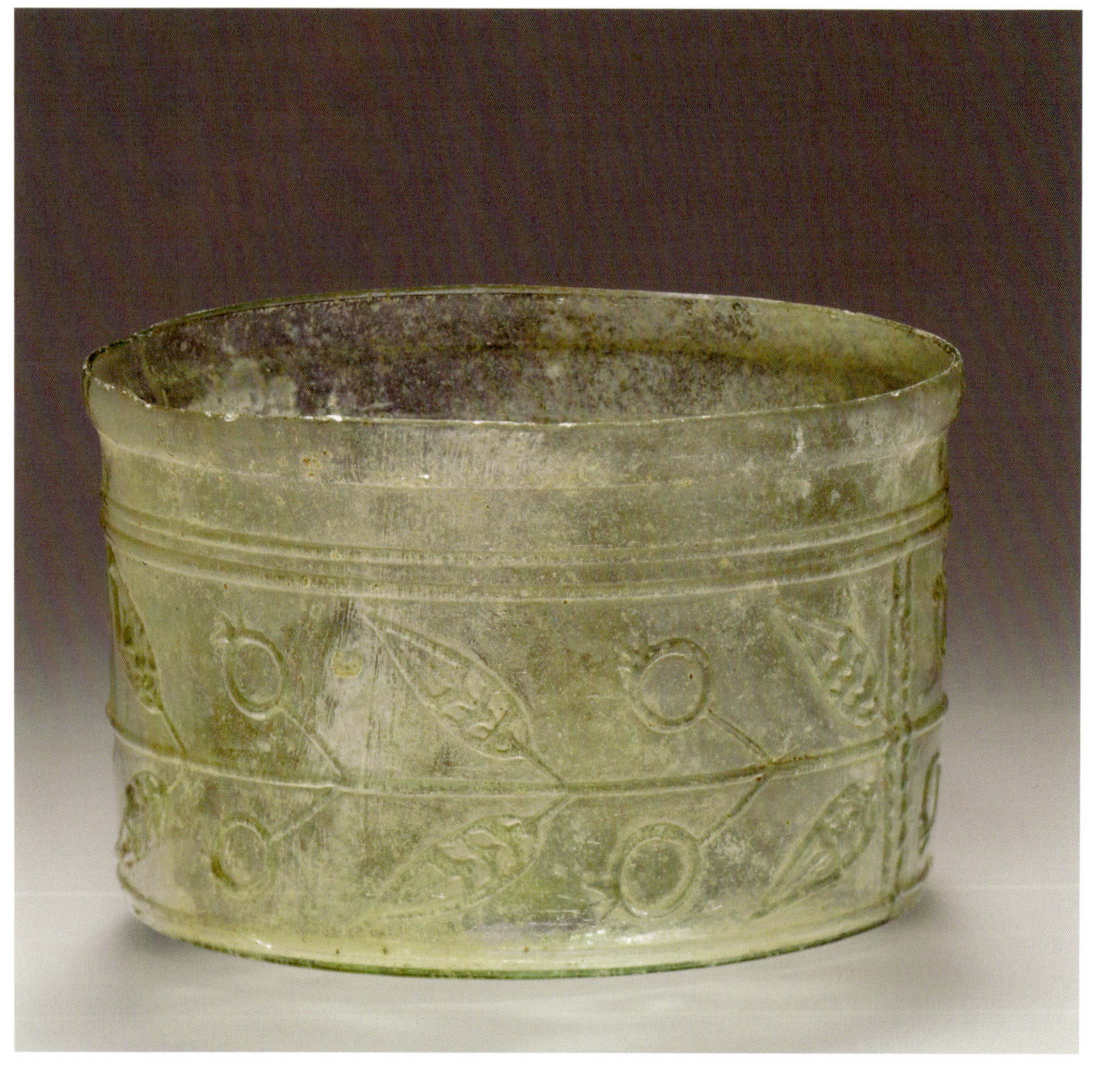

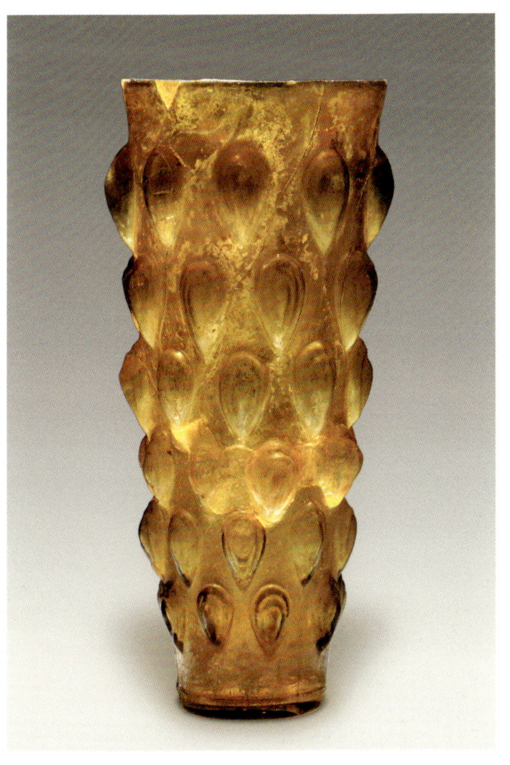

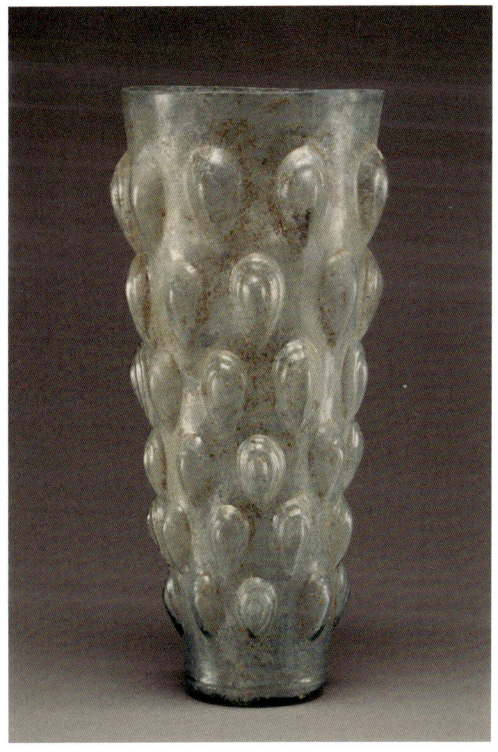

FIGURE 57
Mold-blown cup with pomegranates and grains. Roman, first century A.D. H 5.7 cm
(2¼ in.). JPGM 2003.316.

FIGURE 58
Mold-blown lotus bud beaker. Roman, first century A.D. H 21.2 cm (8⅜ in.).
JPGM 2003.320.

FIGURE 59
Mold-blown lotus bud beaker. Roman, first century A.D. H 21 × DIAM (at rim) 9 cm
(8¼ × 3⁹⁄₁₆ in.). JPGM 85.AF.90.

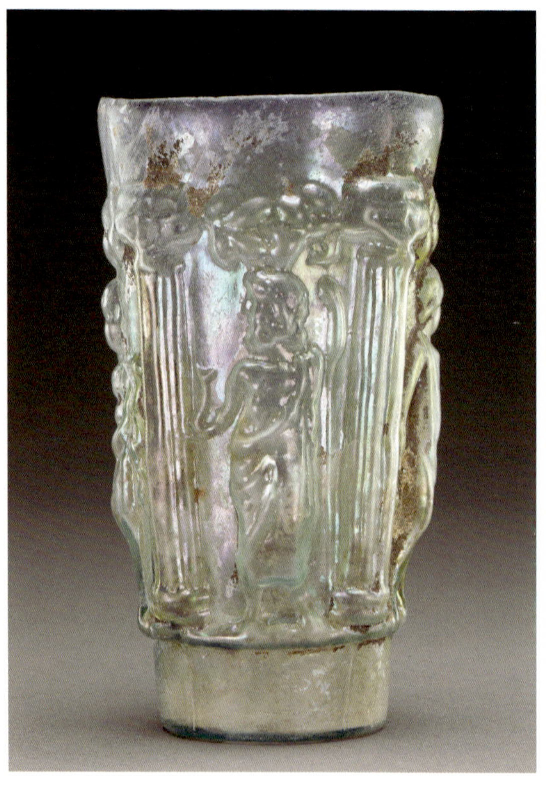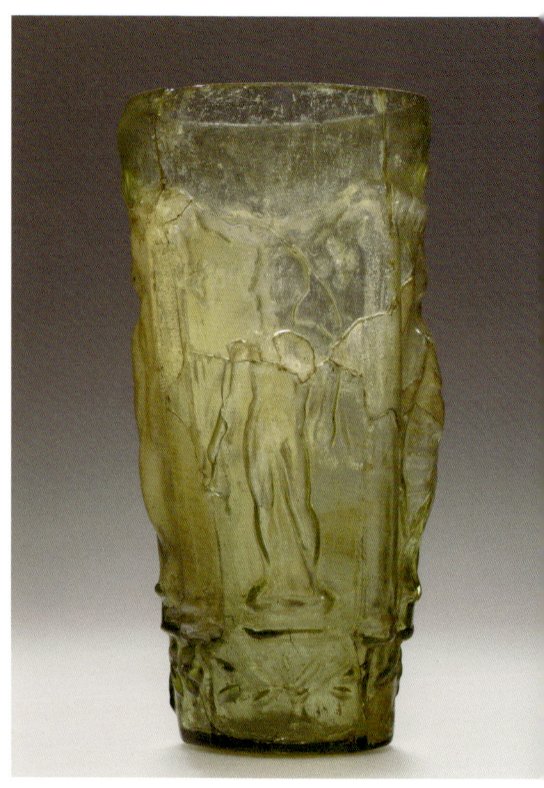

FIGURE 60
Mold-blown mythological beaker. Roman, third quarter of the first century A.D.
H 12.6 × DIAM 7 cm (4^{15}/$_{16}$ × 2^3/$_4$ in.). JPGM 85.AF.83.

FIGURE 61
Mold-blown mythological beaker. Roman, third quarter of the first century A.D.
H 12.5 × DIAM 6.1 cm (4^{15}/$_{16}$ × 2^3/$_8$ in.). JPGM 2003.322.

FIGURE 62
Facet-cut wine cup. Roman, A.D. 75–100. H 8.1 × DIAM (at rim) 6.5 cm (3^3/$_{16}$ × 2^9/$_{16}$ in.).
JPGM 96.AF.320.

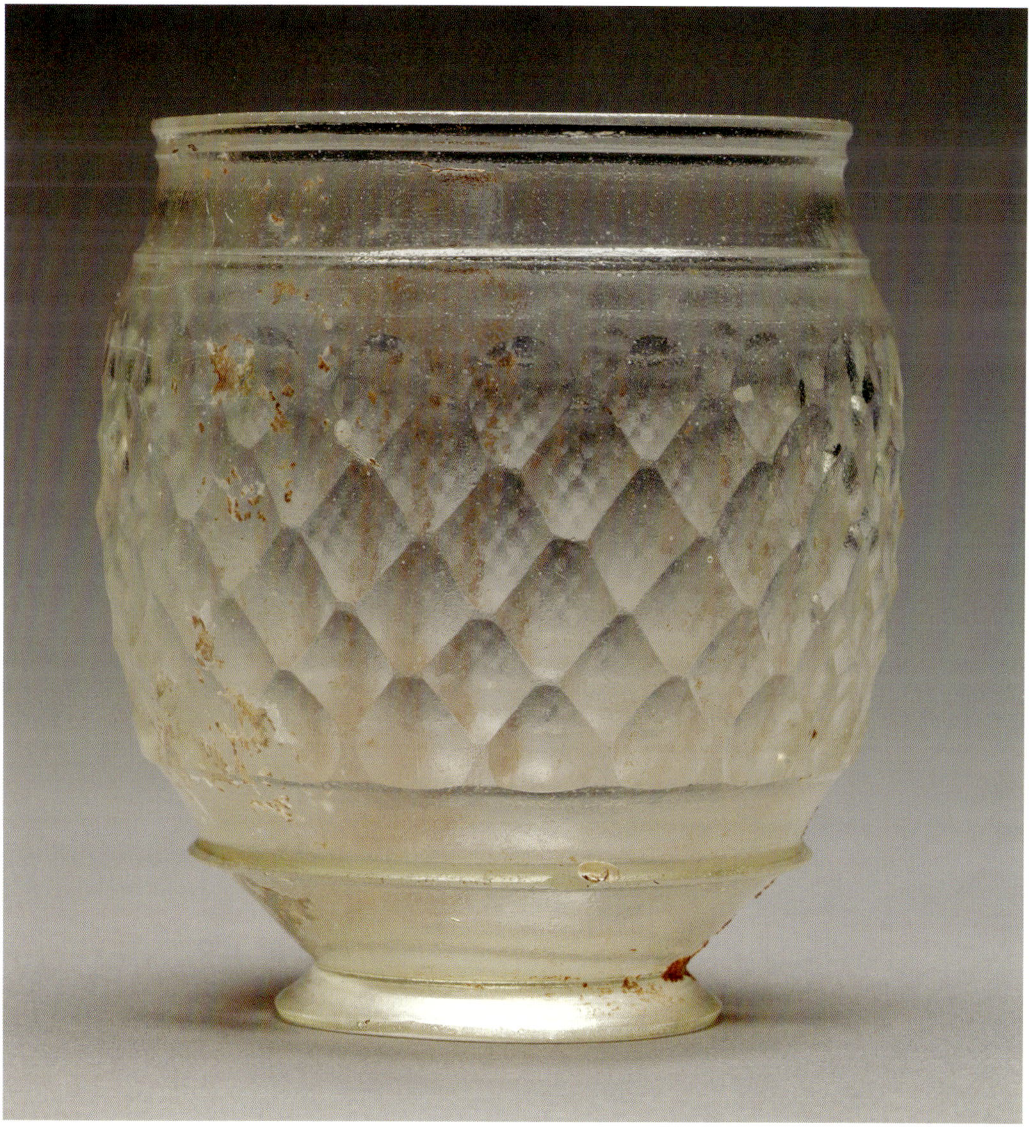

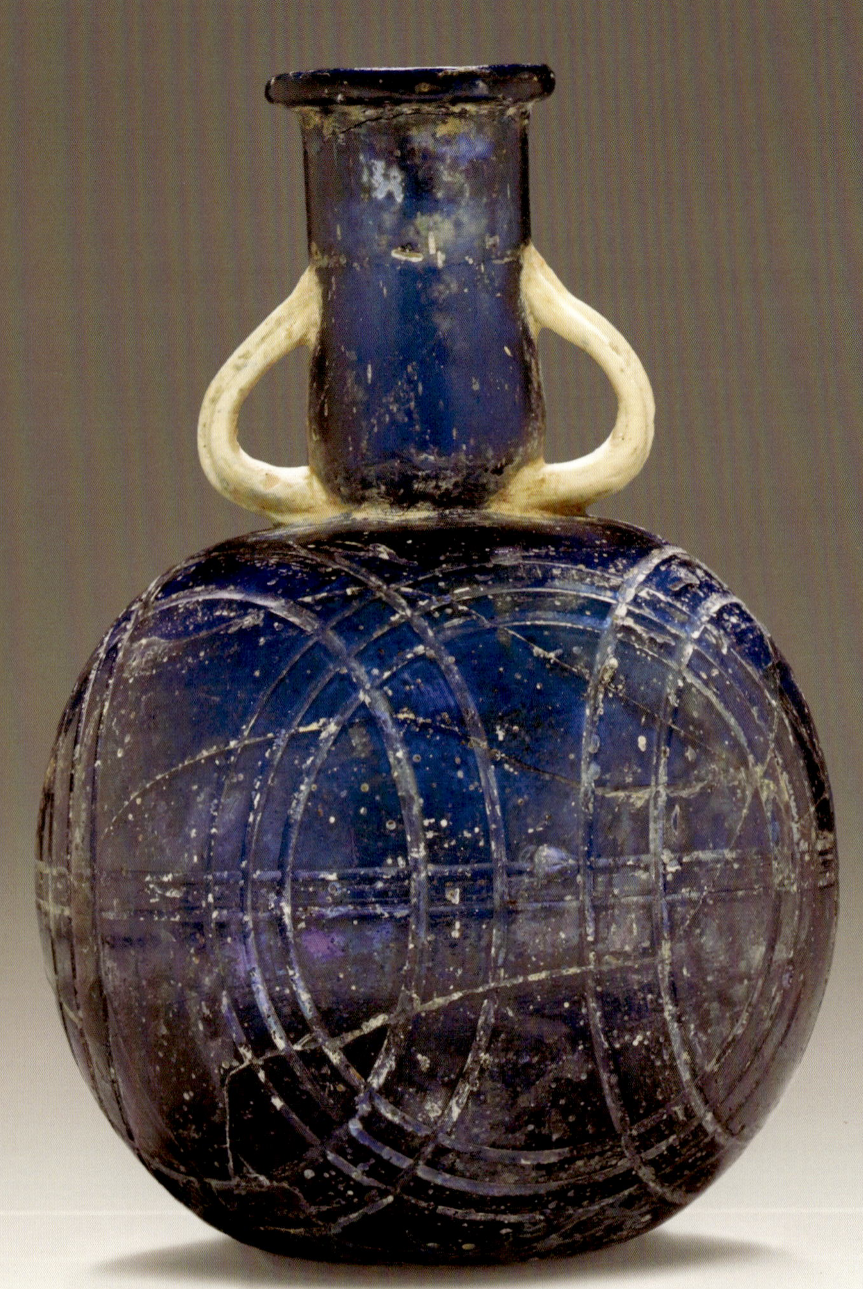

(3¾ × 2¹¹⁄₁₆ in.). JPGM 2003.350.

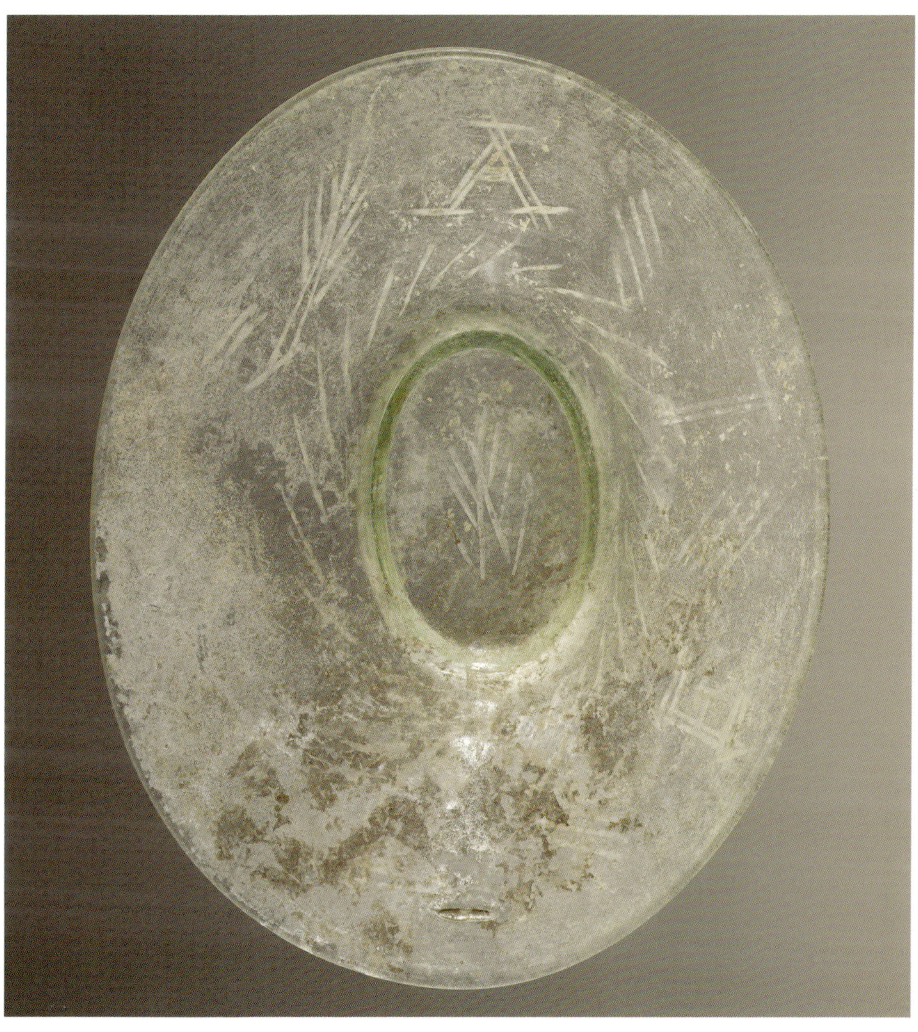

FIGURE 63

Flask with incised decoration. Roman, third–fourth century A.D. H 9.5 × DIAM 6.9 cm
(3¾ × 2¹¹⁄₁₆ in.). JPGM 2003.350.

FIGURE 64

Plate with incised inscription. Roman, third–fourth century A.D. H 4.5 × max. DIAM 23 cm
(1¾ × 9¹⁄₁₆ in.). JPGM 2003.351.

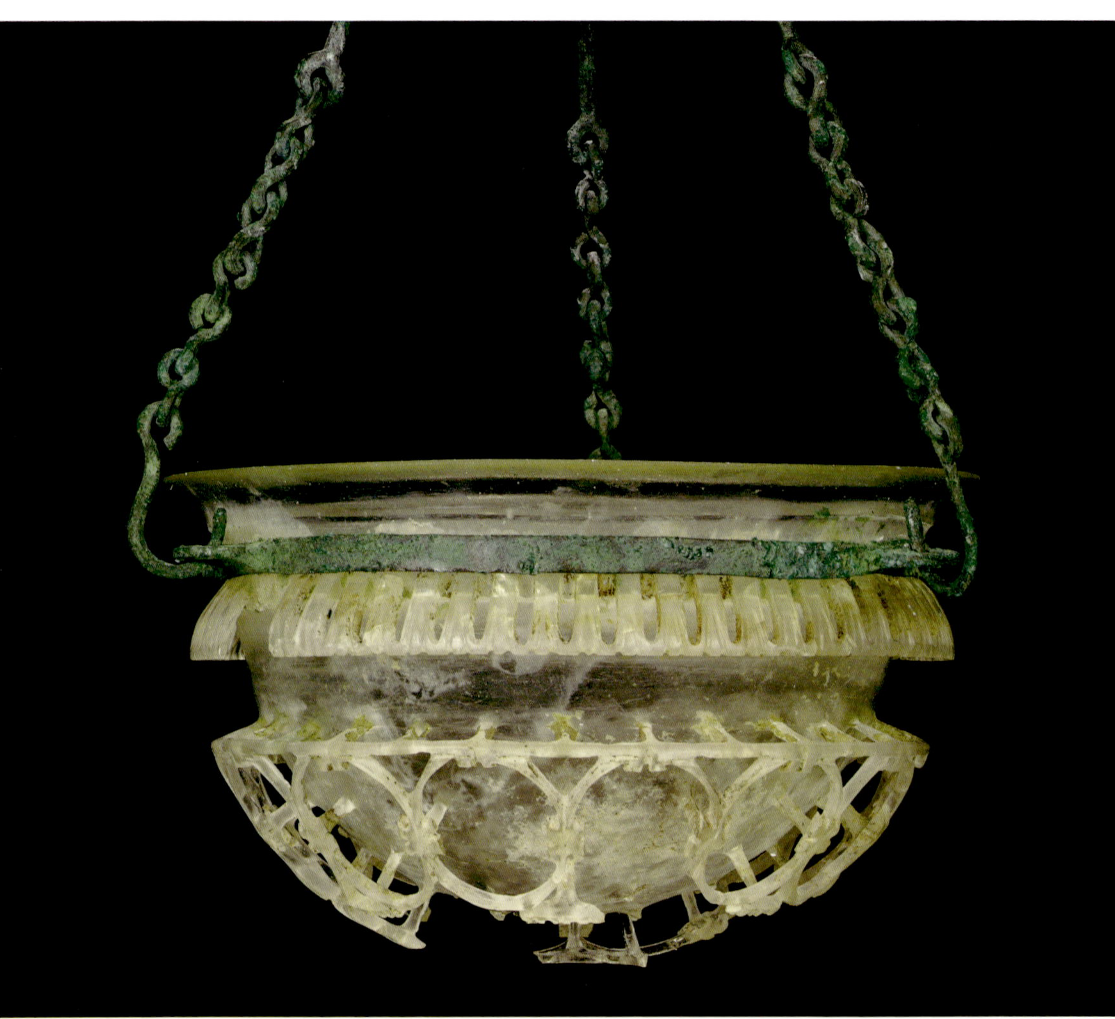

FIGURE 65

Cage cup (diatretum) with suspension chain. Roman, A.D. 300. H 7.4 × DIAM 12.2 cm
(2¹⁵⁄₁₆ × 4¹³⁄₁₆ in.). CMOG 87.1.1. Photo: CMOG.

FIGURE 66

Cage cup (diatretum) with inscription. Roman, first half of the fourth century A.D.
H: 12.1 × DIAM 10.1 cm (4¾ × 4 in.). RGM 60.1. Photo: RGM.

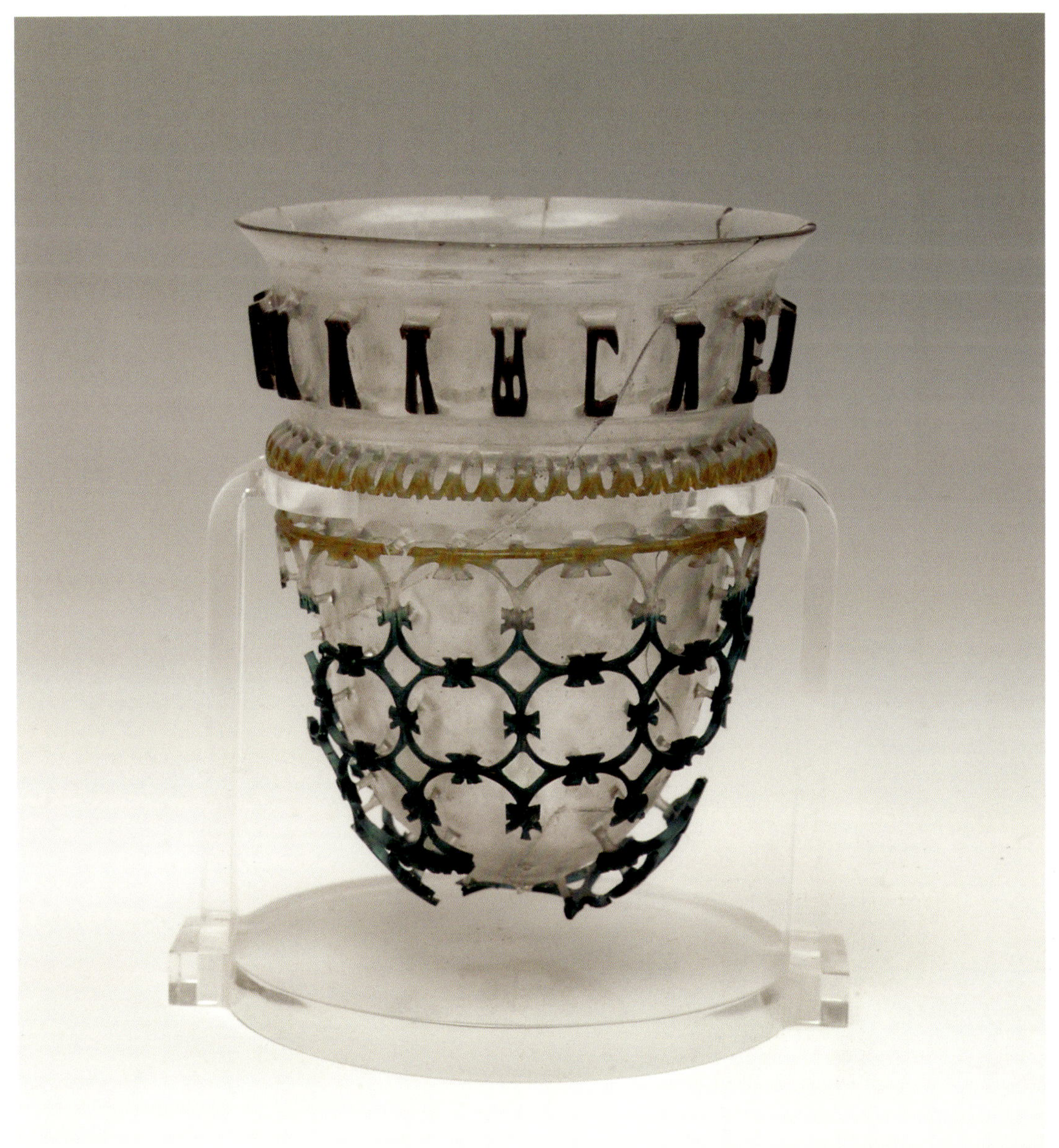

CHAPTER 6

During the Julio-Claudian era, the trade routes for both unworked glass and completed vessels, and also for fragments of glass destined to be recycled, were well established and stretched from Syria and Judaea in the east to the shores of Hispania in the west. As the peace and prosperity of the provinces increased, so did the population, which expanded also the need for manufactured goods, including glass. The glass industry followed this economic development northward from Italy into Gaul, Germania, and even Britannia. Production had always continued in the eastern empire, and as the industry expanded, regional distinctions in the products began to emerge. These regional tastes can be seen most clearly in types whose shapes were so useful they were manufactured everywhere—

Glass
in the
Later Empire
Regional Distinctions

flasks for scented oils, bottles for liquid commodities such as olive oil and wine, sets of dinnerware—but were decorated differently depending upon the local fashion. Some of the regional wares are discussed below in order to provide some sense of how diverse and unique some of them were.

A series of cups known as *Rippenschale* (ribbed cups) have been found in northern Italy and Switzerland. They were manufactured during the first half of the first century A.D. in a multistep process. First, a bubble of glass was formed that would become the base color. Around it were wound white trails of glass, after which the bubble was inserted into a mold that imparted relief ribs to the exterior (Figure 67). The colors used for Rippenschale are saturated and jewel-like, and we can only imagine what a pleasure it must have been to sip wine from such delicate and tactile containers.

Regional distinctions can be seen most clearly in the mold-blown wares, either in the decorative patterns or in the languages used on those with inscriptions. Across the empire, sets of cylindrical drinking cups were designed that carried scenes of gladiatorial combat or chariot racing in the Circus Maximus in Rome or in similar arenas elsewhere throughout the empire. They were the ancient equivalent of the kind of cup we might buy today when attending a sporting event—a plastic container (made in a mold!) for soda or beer that carries images of the athletes and their teams. The same intent was behind the design of the ancient sports cups: pairs of gladiators identified by name face off in combat, or chariots race around the turn posts of the ancient arenas, with their individual or team names above the figures. You can imagine the excitement of the fans and their desire to have some memento of the games. The sports cups that were produced in the eastern empire, where Greek was the common language, contain Greek inscriptions, while those produced in Italy and other western cities have inscriptions in Latin. The majority of the surviving examples are from northern Europe, thus more Latin works are preserved. It is remarkable to see how the names of the figures on the glass cups can be linked to the names of famous combatants recorded by the writers of the day. The name of the gladiator Peitrates, for instance, is preserved both on glass vessels and on an inscription from the city of Pompeii, where he must have fought. The Roman historian Suetonius (Gaius Suetonius Tranquillus, ca. A.D. 69/75–after 130), mentions Columbus, whose name appears on glass vessels (*Lives of the Caesars, Caligula* 55). On the cups with chariot racing, it is the names of the teams that are preserved; interestingly, they are usually colors—the greens, the reds—rather than the names used for sports teams today, which are often fierce predators or warriors.

A similar kind of distinction determined by mold design or language can be found on a series of pitchers and flasks that were purchased by pilgrims traveling to sites of religious significance. Starting in the sixth century A.D., the most popular pilgrimage sites were those directly associated with the life of Christ, in particular, Jerusalem. These pilgrimage vessels tend to take two shapes: small double-sided flasks, or larger hexagonal bottles or pitchers that have rudimentary images of geometric designs and religious symbols alternating around the body of the vessel. An example in the Getty collection illustrates the shape, but its molded decoration contains no religious symbols (Figure 68).

One type of pitcher includes three variant images of the True Cross alternating with diamond-shaped devices. Examples made from this mold design have been found in Damascus, Tyre, Gerasa, Beth She'arim, and Nazareth. These vessels all contained water or oil that could be blessed and then brought home as a souvenir. The glassmakers did not restrict themselves to Christian customers alone, and the exact same shape of vessel with a hexagonal body was decorated with three Jewish motifs, including the menorah, separated by the same diamond-shaped elements. Similar religious souvenirs can be found across the empire as the worship of Christianity spread. As churches were built to honor particular saints and to house their remains or to mark the site of their martyrdom (St. Peter's Basilica in Rome, for example), the pious would undertake strenuous travel in order to honor the saint. Just as the gods and goddesses of ancient Greece and Rome had their specific attributes, the saints were associated with specific symbols that glassmakers then used in their designs of flasks for holy water or oil. Military conflict and political unrest directly affected the waxing and waning of religious tourism in different parts of the empire, and the glass industry had to adapt to the constant changes of supply and demand.

Regional style can also be identified by the preponderance of a specific shape being found in only one part of the empire. This is true of the glassware made in Colonia Claudia Ara Agripinnensium, present-day Cologne. The city is strategically located on the Rhine River, thus it was an important place to locate Roman troops before the area was fully settled. It was the birthplace of Agrippina the Younger, wife of the emperor Claudius, whose father, Germanicus, had made it the headquarters of his military campaigns. The city was elevated to the status of a *colonia*, a city under Roman law, in A.D. 50, and became the provincial capital in A.D. 90. One variety of decorative glass associated with the city is known as snake thread, as the trails of hot glass that were wound over the surface of the vessel resemble the sinuous movement of snakes. The ends of these trails were often flattened on the surface of the vessel using tools that imparted a pattern that provided variety to the thickness and shape of the applied glass design. They were made during the third and fourth centuries A.D. The standard variety is a combination of a colorless bottle with either colorless or colored threads (Figure 69). The bottles came in a variety of shapes, from small perfume flasks to large wine carafes. When colored threads are used, they are often a combination of opaque white and opaque blue (see Figure 41 for an eastern variant with white and blue threads). Another type of glass vessel associated with Cologne had applied decorative elements such as fish and dolphins. These added elements were made separately before the vessel was formed and then attached to the exterior as high-relief decoration (Figure 70).

Some stylistic distinctions are made not by form or decoration, but by subsidiary elements such as the manner in which a handle or foot is added to a vessel. Such is the case with a group of vessels that share a common characteristic in their handle design. The Getty Museum is fortunate to have two examples of this workshop's production in its collection (Figures 71, 72). What makes this group interesting is that a variety of shapes were made using this handle design, both mold blown and free blown. The two examples in the Getty collection are mold-blown head flasks. Both were made using cobalt blue glass; three other examples of such blue glass head flasks are known to have survived from antiquity, as well as one clear glass example made from the same mold. After the head shape had been formed in the mold, it was removed, and a thick trail of glass was wound around the bottom to form a ring-shaped base. Next, a handle was attached by taking a thick gather of glass, attaching it to the neck of the flask about halfway down, and then attaching the other end of the handle on the back of the head. The handle was then pulled away from the vessel, and a tool was used to make a large pinch in the glass to form a thumb rest. The glass at the base attachment was then stretched down the back of the head, and a series of horizontal ridges was formed by using a pincer or tongs to pinch folds into the glass. This distinctive handle shape is sometimes called a wishbone handle because of the pinch halfway down its length. The same handle appears on a small number of mold-made pear-shaped and globular pitchers and bottles, also composed of cobalt blue glass, that appear to have been created in the eastern Mediterranean region sometime between the mid-fourth and mid-fifth centuries A.D.

Our understanding of regional styles and distinctions is almost completely reliant on the archaeological record. Types are first identified from these excavated finds, and relationships are then made in collections that have similar examples but no provenance. As excavations continue, it is presumed that even more distinctive types will be identified to broaden further our understanding of how the glassmaking industry developed its products throughout the Roman Empire.

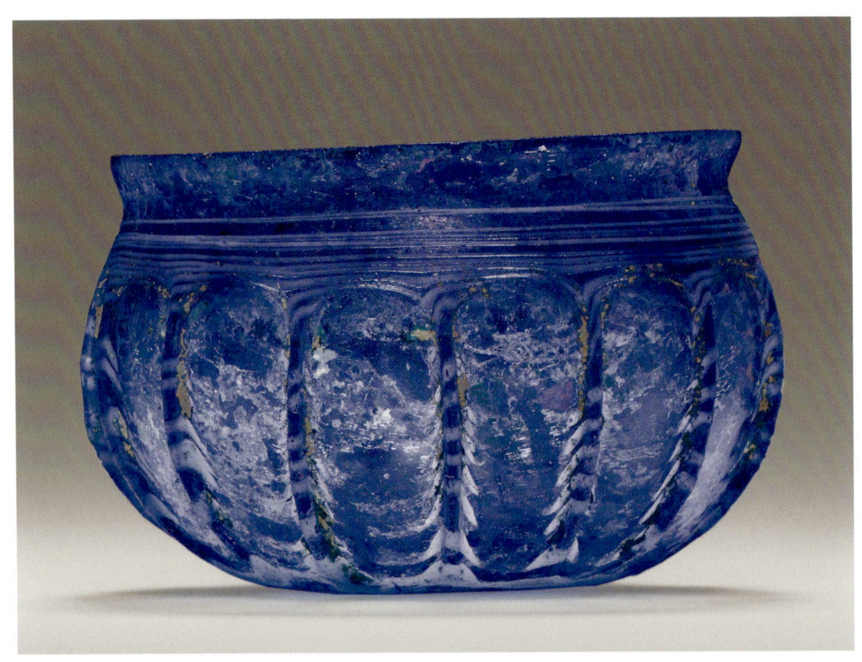

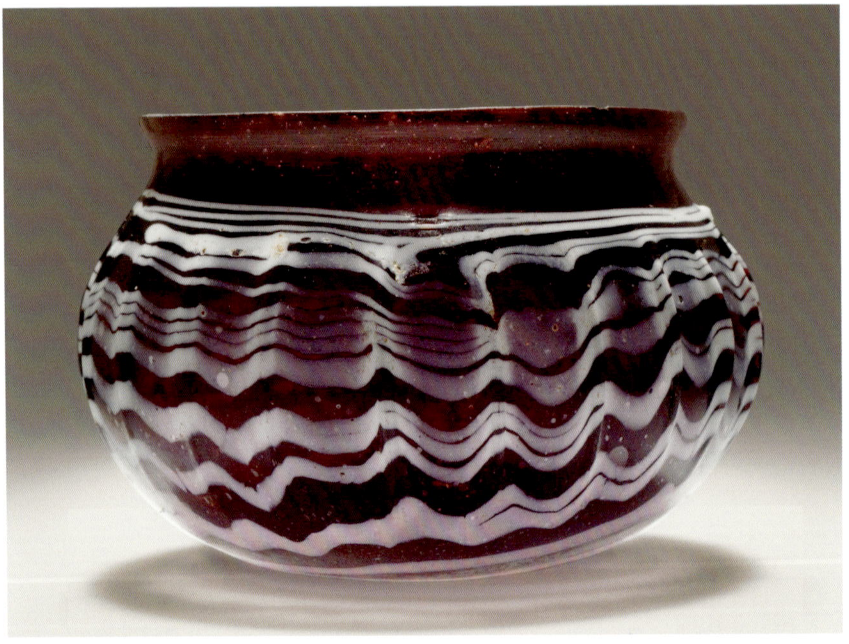

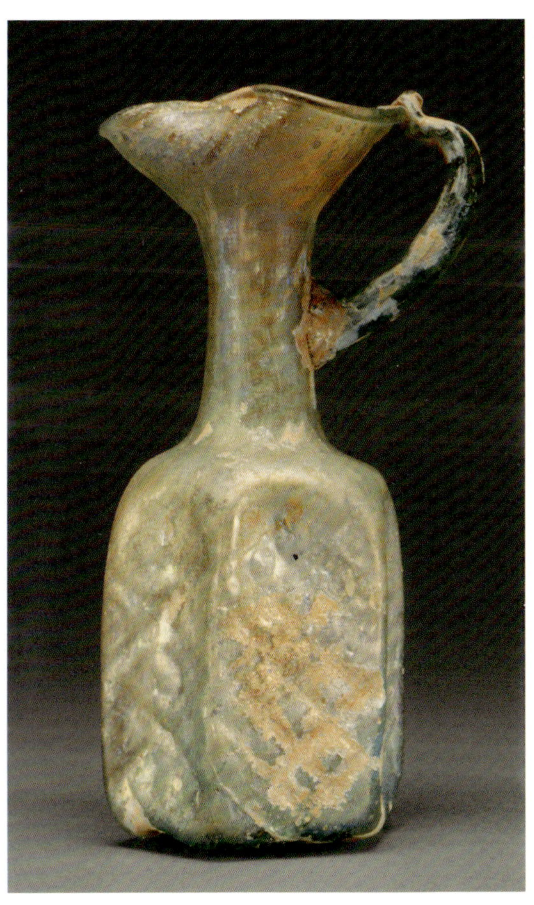 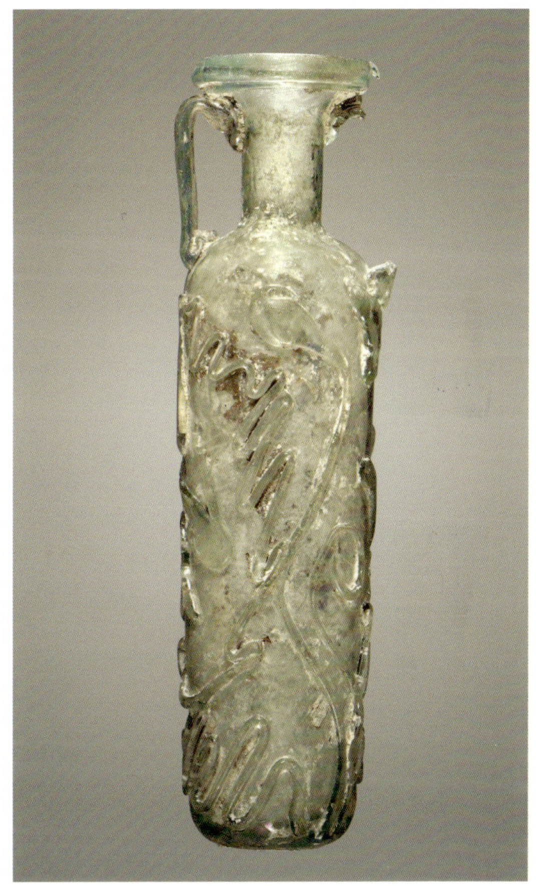

FIGURE 67
Two *Rippenschale*. Roman, second half of the first century A.D. (*left*) H 5.8 × DIAM 8.5 cm
(2 5/16 × 3 3/8 in.). JPGM 2003.226, (*right*) H 6.5 × DIAM 8.1 cm (2 9/16 × 3 3/16 in.).
JPGM 2004.20.

FIGURE 68
Mold-blown hexagonal pitcher (oinochoe). Roman, sixth century A.D. H 14 cm (5 1/2 in.).
JPGM 78.AF.24.

FIGURE 69
Snake-thread perfume flask. Roman, third–fourth century A.D. H 11 × DIAM 3.8 cm
(4 5/16 × 1 1/2 in.). JPGM 2003.425.

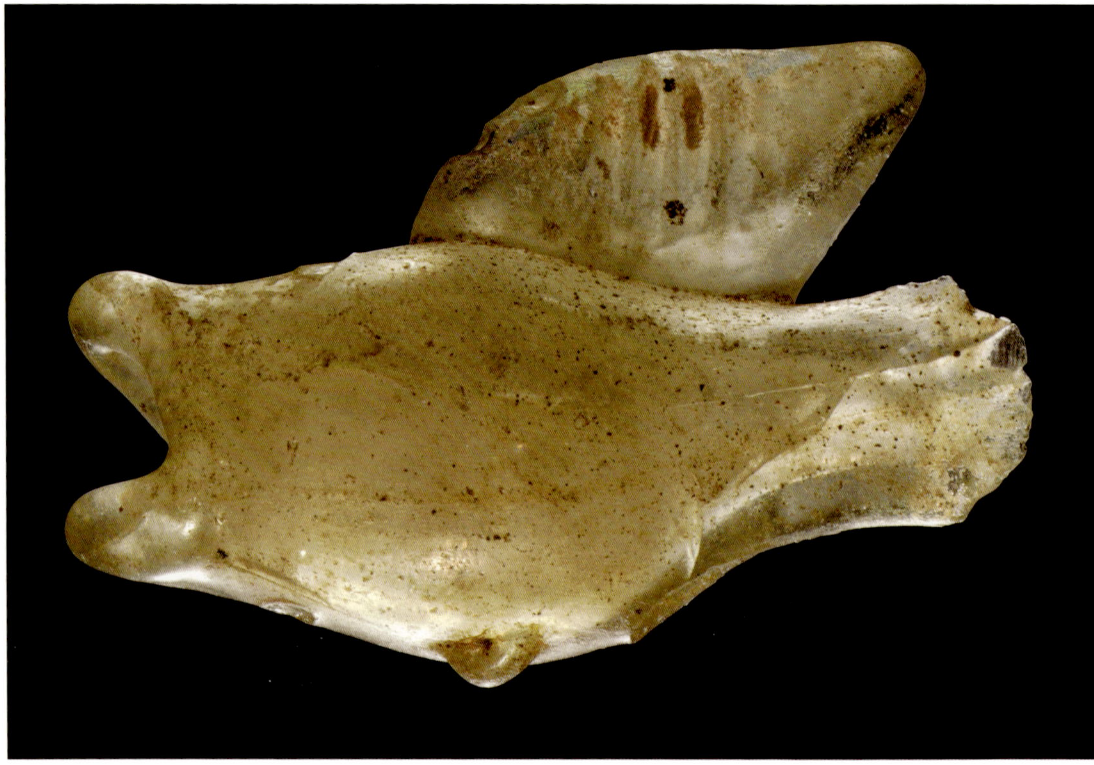

FIGURE 70
Fish-shaped attachment to a vessel. Roman, third century A.D. L 3.5 cm (1⅜ in.).
JPGM 2003.405.

FIGURE 71
Mold-blown head flask. Roman, fourth–fifth century A.D. H 17 cm (6¹¹⁄₁₆ in.).
JPGM 2004.44.

FIGURE 72
Mold-blown head flask. Roman, fourth–fifth century A.D. H 17.2 × W 8 cm (6¾ × 3⅛ in.).
JPGM 85.AF.320.

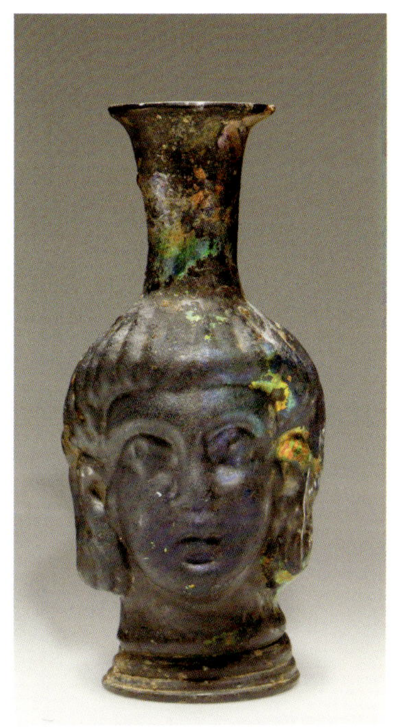

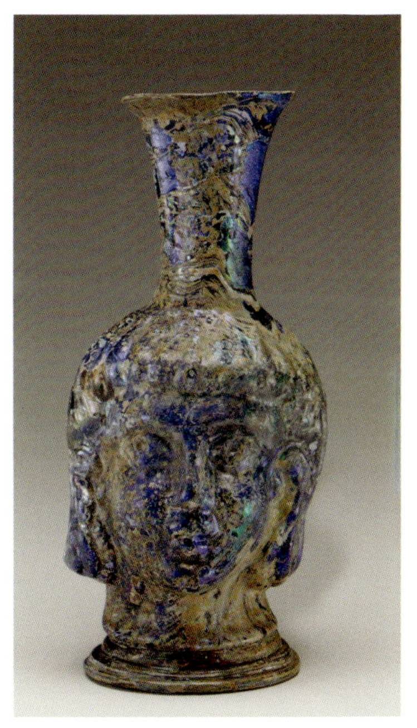

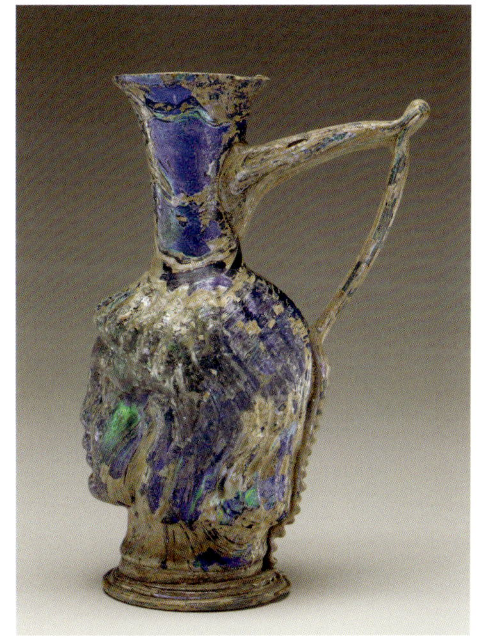

CHAPTER 7

Our survey of glassmaking techniques has sometimes addressed the functions to which glass objects were put in antiquity. It is not difficult to determine how the ancients used glass in their lives since in many instances we use glass the same way. If we look around our homes, we can see glass applied in almost every room. What differentiates us from the ancients, however, is how the glass was made, and how, over twenty-five hundred years, glass changed from a material reserved exclusively for the highest echelons of society to something that was prevalent in every household.

One of the earliest uses of glass was for jewelry, and scholars believe that glassworkers were trying to replicate costly precious stones. The earliest technique

How Was Glass Used?

of casting glass was used to make pendant beads and inlays, as well as small vessels. Glass jewelry remained popular in later periods, and beads continued to be made (Figures 73, 74). A pair of gold glass pendants in the Getty collection is an example of some of the unusual items that took advantage of the unique properties of glass. The pendants have images of birds cut from gold foil that were covered with clear disks of glass and then set into bronze surrounds (Figure 75).

In later periods, cast glass continued to be used for gems and also for small sculptures (Figure 76). A stunning example of a glass sculpture in the Corning Museum of Glass is the torso of Venus, the Roman goddess of love (Figure 77). Although the current surface of the glass has altered to a flesh-colored encrustation, the glass used to make the figure was a beautiful deep green. In the Hellenistic period, the mosaic and casting techniques were combined to create fanciful sculptural figures, including snakes (Figure 78).

Glass was used to embellish architecture in the form of decorative surrounds placed around doors and windows, in column capitals, and in the mosaics that covered floors and ceilings (Figure 79). While the majority of mosaics were made using stone tesserae (small squares), glass was often introduced to provide colors that were not readily available, if at all, in stone. Thus green, blue, and red tesserae were frequently made of glass. The Romans introduced the use of glass in windows. Unlike today's plate glass, Roman windows were cast as small panes that were then inserted into grilles of wood or bronze.

But for the most part, glass was used to manufacture vessels in antiquity. The earliest examples were small perfume flasks made using the mosaic and core-forming techniques, and many of these vessels were suspended on chains to be worn attached to a belt at the waist or around the wrist. The Getty Museum has examples with their bronze chains still attached (Figure 80). Before and after the discovery of glassblowing, the shapes of these early perfume containers mimicked their cousins in both ceramic and metal or took on the more fanciful shapes that the medium allowed, such as a series of perfume flasks with amusing tripod feet (Figure 81). Such smaller containers were also used to hold cosmetics (Figure 82). An ingenious design was developed for holding kohl (used as an eye cosmetic). Often called kohl tubes, they were formed by taking a hollow tube of glass and folding it in half, thus neatly creating two tubular containers for the precious material. One example in the Getty Museum still has a bronze applicator rod inside (Figure 83). Rods of glass were also made to dip into these elongated containers to remove their contents (Figure 84).

With the invention of glassblowing in the first century B.C., glass vessels began to replace ceramic wares in the kitchen and for dining. Pitchers were made to serve water or wine, and small cups to drink these beverages (Figures 85, 86). One unusual cup assemblage in the Getty collection is composed of a vessel shaped as a miniature krater, a form associated with the mixing of wine and water at banquets. This mixing was necessary, because the alcohol content in ancient wine was higher than it is today, and taking wine without diluting it would lead quickly to inebriation. Nestled inside its rim is a smaller cup. Perhaps this was an individually sized mixing set so that water and wine could be mixed in the miniature krater, and then poured into the smaller cup for consumption. Sets of dishes were made to serve the food at table, to receive the food, and to present the various sauces and condiments that accompanied each dish (Figures 87, 88). The mold-blowing technique allowed for the creation of exact duplicates, and the mold designs spoke directly to the drinker or diner. Two small wine cups at the Getty have a mold design with a Greek inscription that tells the holder ΕΥΦΡΑΙΝΟΥ ΕΦW ΠΑΡΕΙ ("Be glad that you have come") (Figures 89, 90). The measuring cups used to prepare dishes were made of glass, and even serving and eating utensils could be made of the material (Figure 91).

Other household uses for glass included illumination. Glass was the material utilized to make lamps that resembled their counterparts in clay, as well as cup-shaped oil holders that were placed in chandeliers (Figures 92, 93).

Unique containers were also designed to dispense small amounts of liquid, possibly to refill oil lamps, and had extremely narrow spouts that emerged from the bulbous body of the container (Figure 94). These spouted vessels are sometimes called "feeder" flasks since some were thought to have been used as milk bottles for infants. Inkwells also were made of glass (Figure 95).

Glass even served a purpose in the afterlife, as cinerary urns were made to hold the bones and ashes of the deceased (Figure 96). Glass vessels and pieces of jewelry were also left in the grave as gifts for the deceased. Many of the intact vessels now in museums around the world have come from graves, and the soil conditions in which the glass was buried have directly affected its appearance.

The two head flasks described in chapter 6 for their regional distinctiveness are a perfect pair to illustrate an element of ancient glass that has often entranced modern admirers (see Figures 71, 72). These two heads, having been manufactured in the same mold, are nearly identical in shape, but not in appearance. The flask shown in Figure 72 is weathered and iridescent, while the other appears to be in better condition—differences that resulted from weathering and aging and the conditions in which the vessels were buried. As glass ages, a chemical reaction caused by environmental conditions occurs on its surface. Known as weathering, this reaction often involves the leaching of alkali from the glass by water. As the glass delaminates, thin layers are formed on the surface. In some instances, these layers take on an iridescent sheen, but in others, they simply form an unattractive crust. The outcome is purely coincidental, thus the iridescence that is so prized by collectors today is a fluke. The eighteenth- and nineteenth-century rediscovery of ancient glass objects with iridized surfaces led modern glassmakers such as Louis Comfort Tiffany and Émile Gallé to develop manufacturing techniques that intentionally re-create these ancient surfaces, and they certainly achieved stunning results. Modern iridescence is accomplished either by introducing metallic substances into the batch or by spraying the surface of the glass with stannous chloride or lead chloride while the vessel is being constructed, and then reheating it in an oxygen-deprived atmosphere in the kiln.

The craft of glassmaking has continued uninterrupted from the second millennium B.C. to the present day and is now considered an art form. Modern technology and scientific advancement have enabled contemporary glassmakers to expand the applications of glass, and the techniques used to make glass will undoubtedly continue to evolve as both artists and scientists continue to experiment with this changeable and pliant medium. But it is important to remember that the basic methods for fashioning glass into usable objects have remained largely unchanged for two millennia, and we owe much to the ingenious artists who created, manipulated, and transformed this fascinating material. One only need look at a bowl from the Roman period in the Corning Museum of Glass that was designed with an undulating rim, and compare it to one of Dale Chihuly's *Sea Forms*. The Roman vase is the only one known of this design, and the craftsman who made it first fashioned a simple bowl, and then carefully reheated the edge, allowing it to deform so he could shape it into a rippling rim. Modern Italian glassmakers working at the Venini studios produced their own handkerchief vases without any knowledge of this piece, which was acquired by the Corning Museum only in 1997. Inspired by the modern Venini creations, Dale Chihuly created a similar design with rippling edges in his *Sea Forms*. But unlike the Roman glass, the Venini and Chihuly pieces were not generated for functional reasons; rather, they were made by studio artists as works of art.

It is humbling to realize that the Romans have always beaten us to the punch when it comes to creativity and glass. And it is a testament to the medium that glassblowers working two millennia apart were able to take the same material, understand its complexities and frailties, and create the same kind of beautiful object. All glass artists working today have a deep respect for their ancient

Roman counterparts and what they managed to accomplish with the limited technology and scientific knowledge available to them. Corning's fourth-century bowl and Chihuly's *Sea Forms* are, in the end, essentially the same silica, alkali, and lime, manipulated in the hands of masters into unique creations that will stand the test of time.

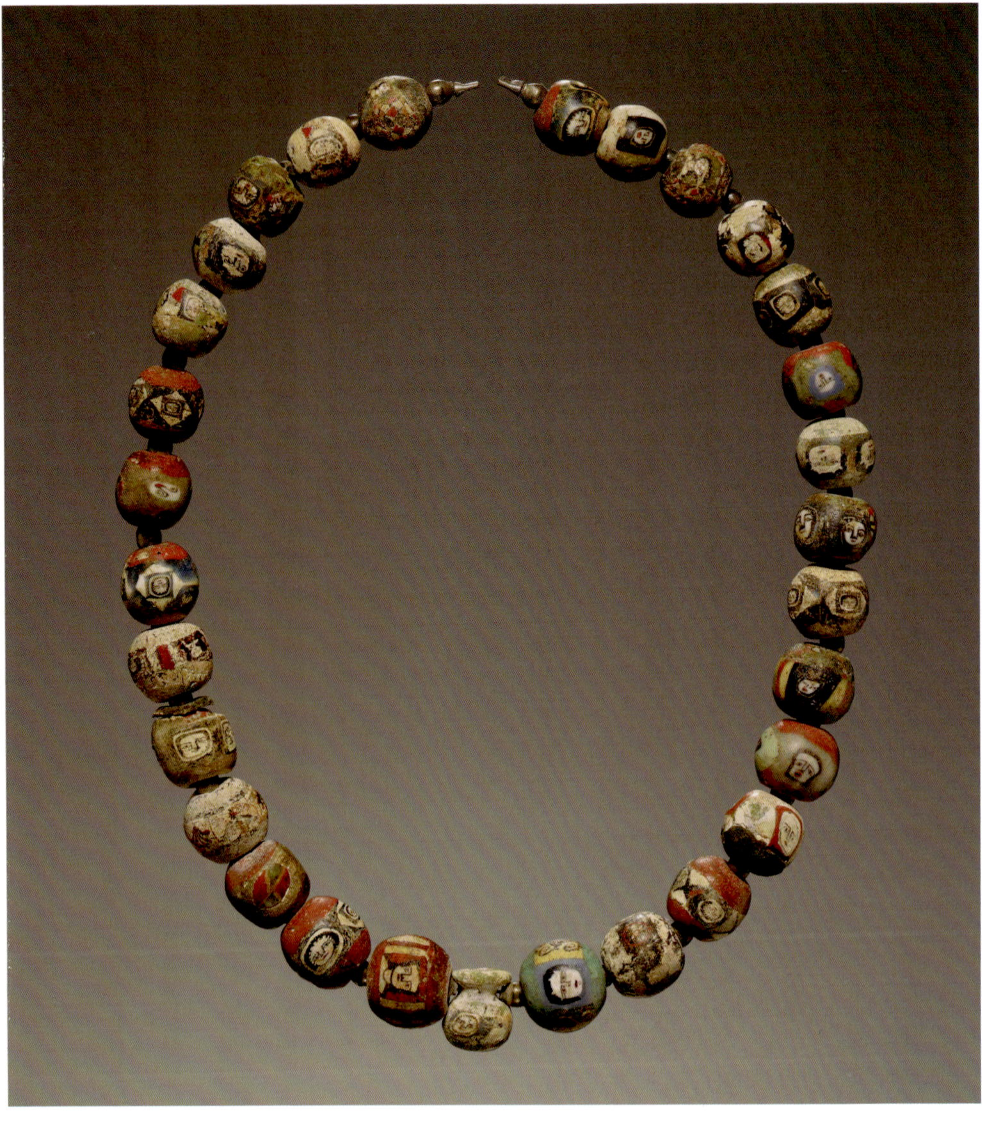

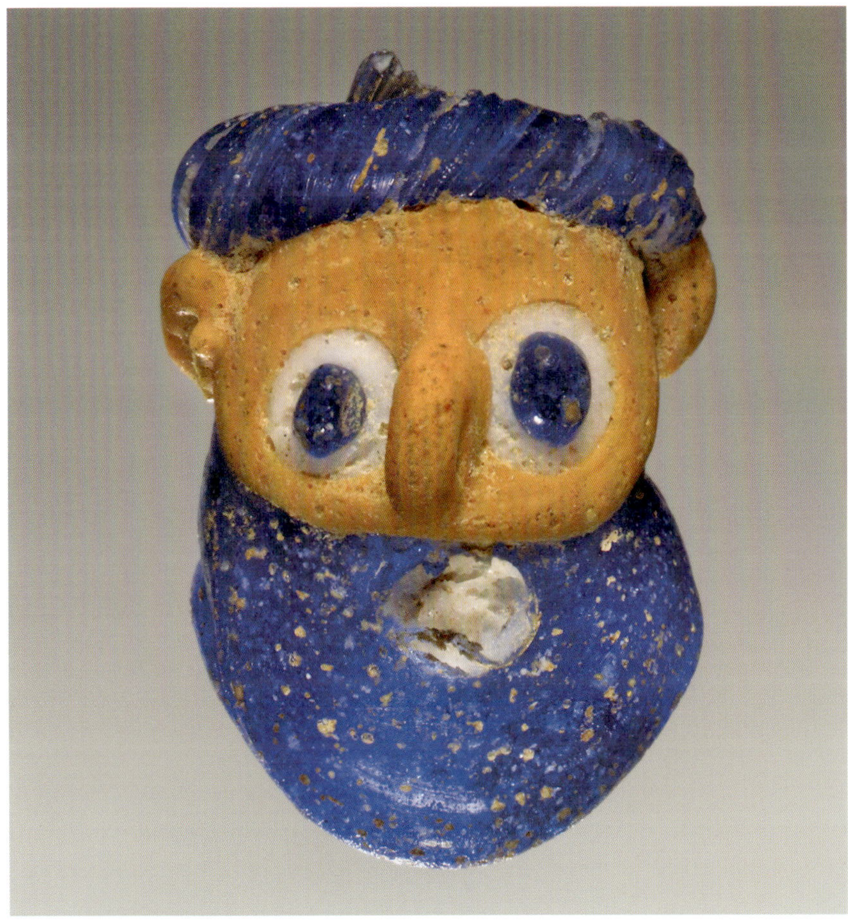

FIGURE 73
Necklace composed of thirty mosaic glass beads. Roman, first century B.C.–first century A.D.
DIAM 1.2–1.7 cm ($\frac{1}{2}$–$\frac{11}{16}$ in.). JPGM 2003.259.

FIGURE 74
Head bead. Greek, fifth–fourth century B.C. L 1.8 cm ($\frac{11}{16}$ in.). JPGM 2003.207.

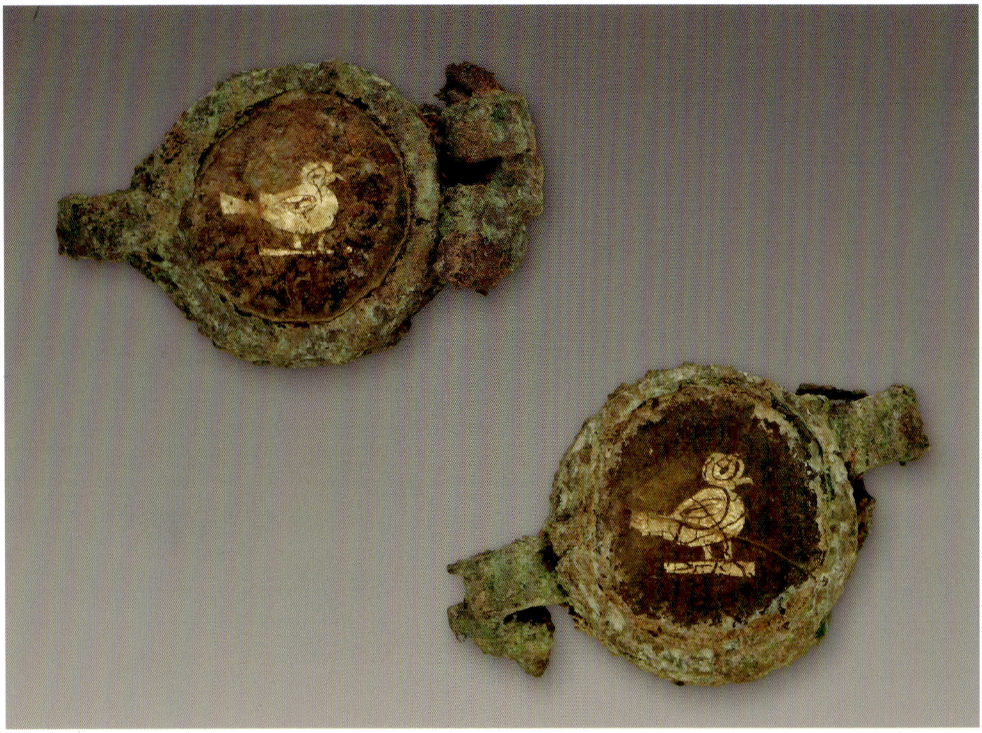

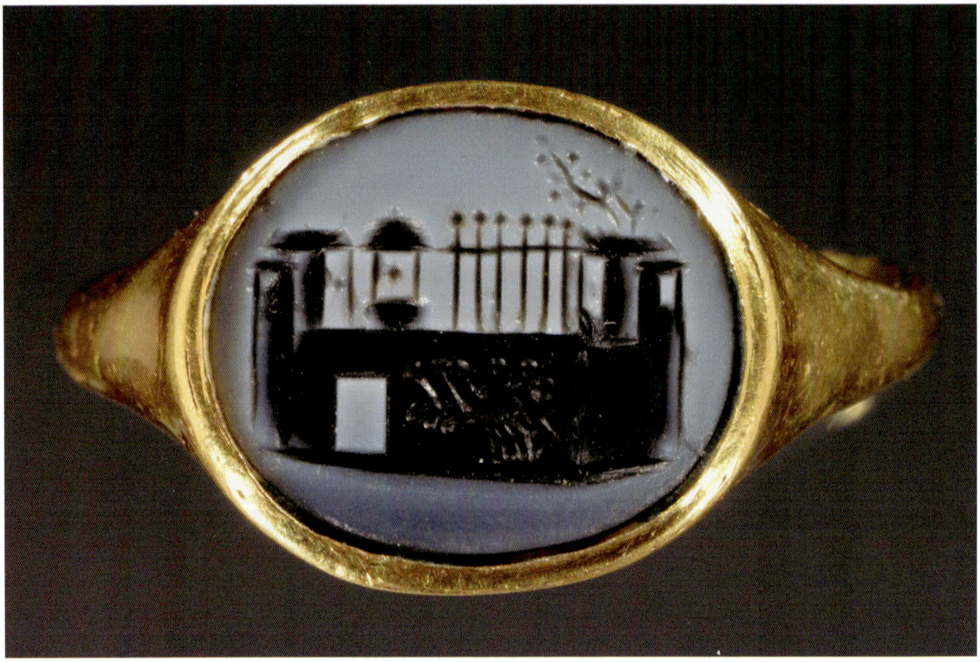

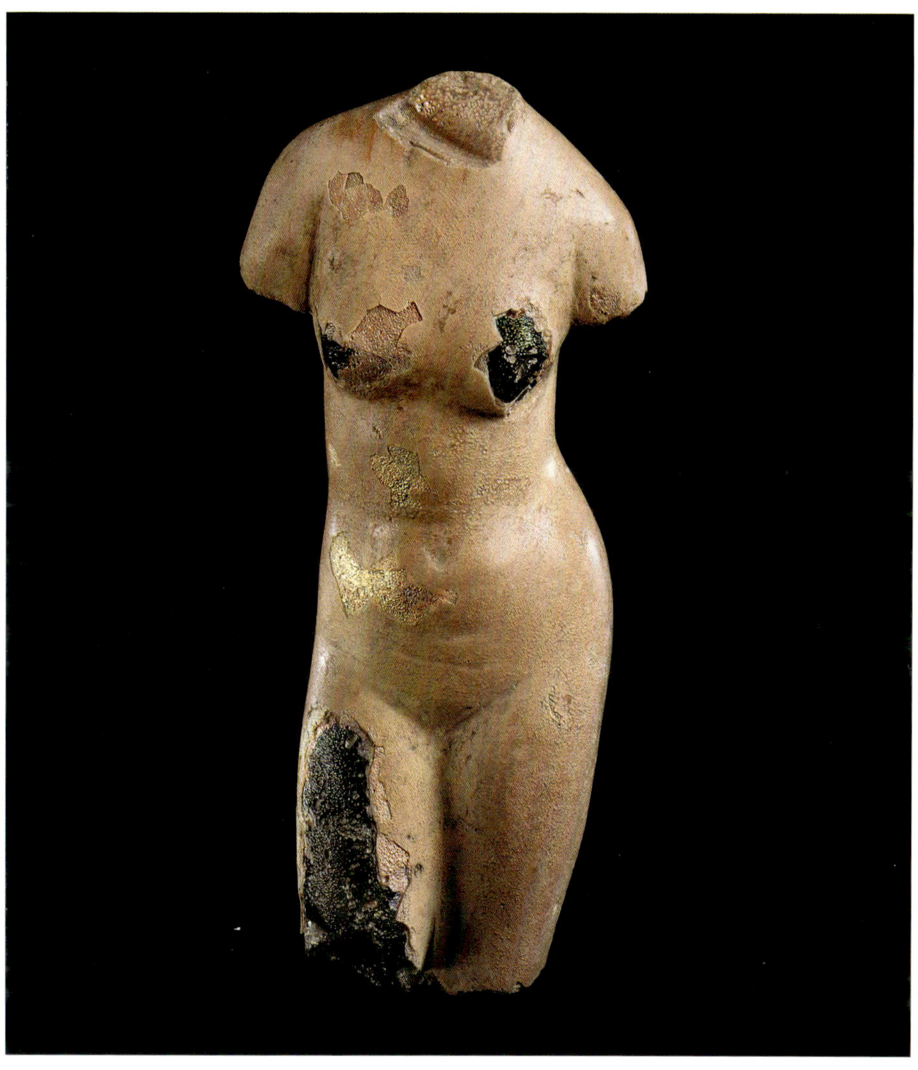

FIGURE 75
Pair of bronze pendants with gold glass inlays. Roman, first–second century A.D. Each DIAM 1.1 cm
(⁷⁄₁₆ in.). JPGM 2003.295.

FIGURE 76
Gold ring with an inlaid glass cameo depicting Achilles dragging Hector's body before the walls
of Troy. Roman, first century A.D. Ring (approx.): H 1.2 × W 1 cm (¹⁄₂ × ³⁄₈ in.). Bezel (max.):
DIAM 2.1 cm (¹³⁄₁₆ in.). JPGM 84.AN.857.

FIGURE 77
Cast torso of Venus. Roman, first–second century A.D. H 9.4 × W 4.5 cm (3¹¹⁄₁₆ × 1³⁄₄ in.).
CMOG 55.1.84. Photo: CMOG.

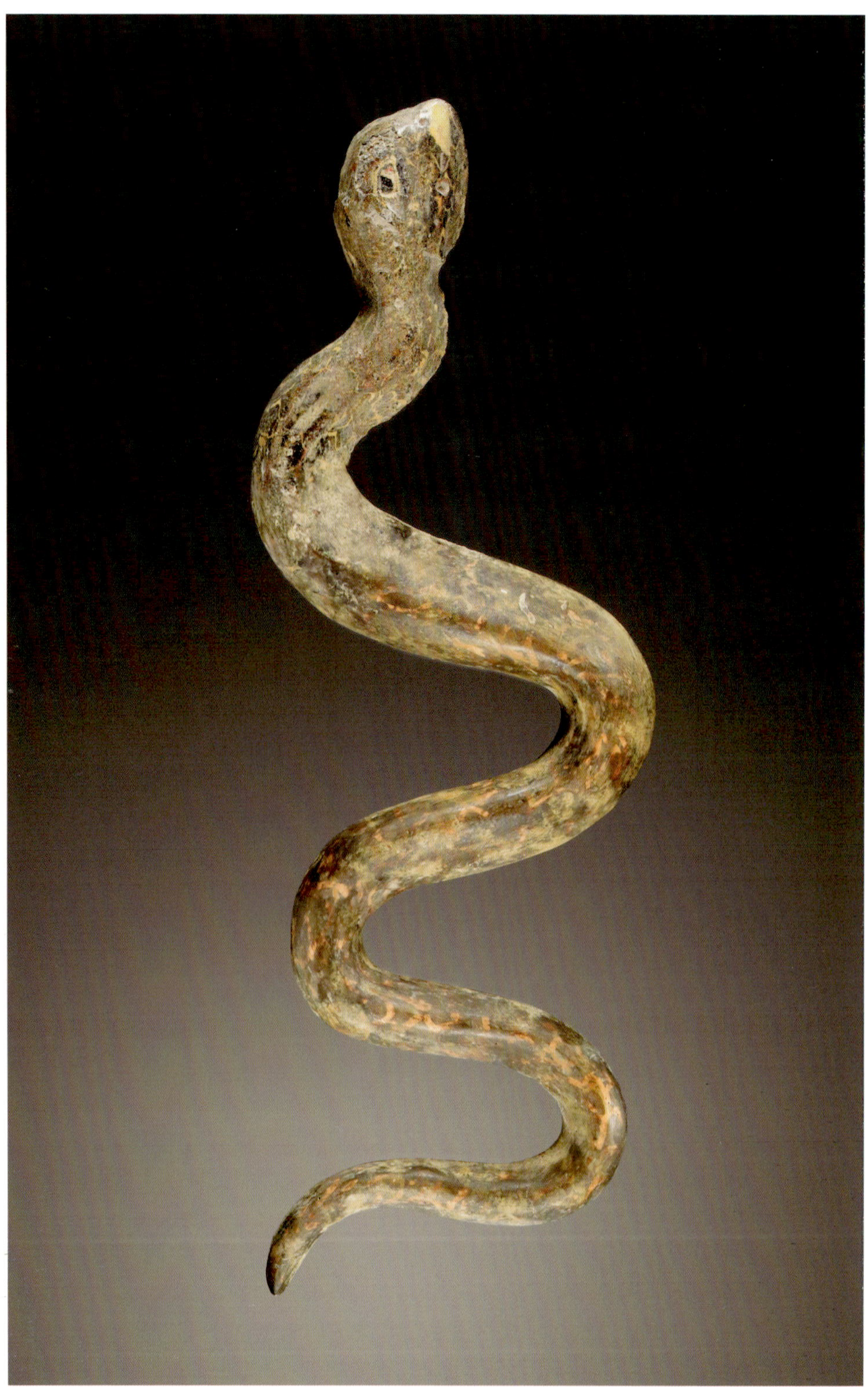

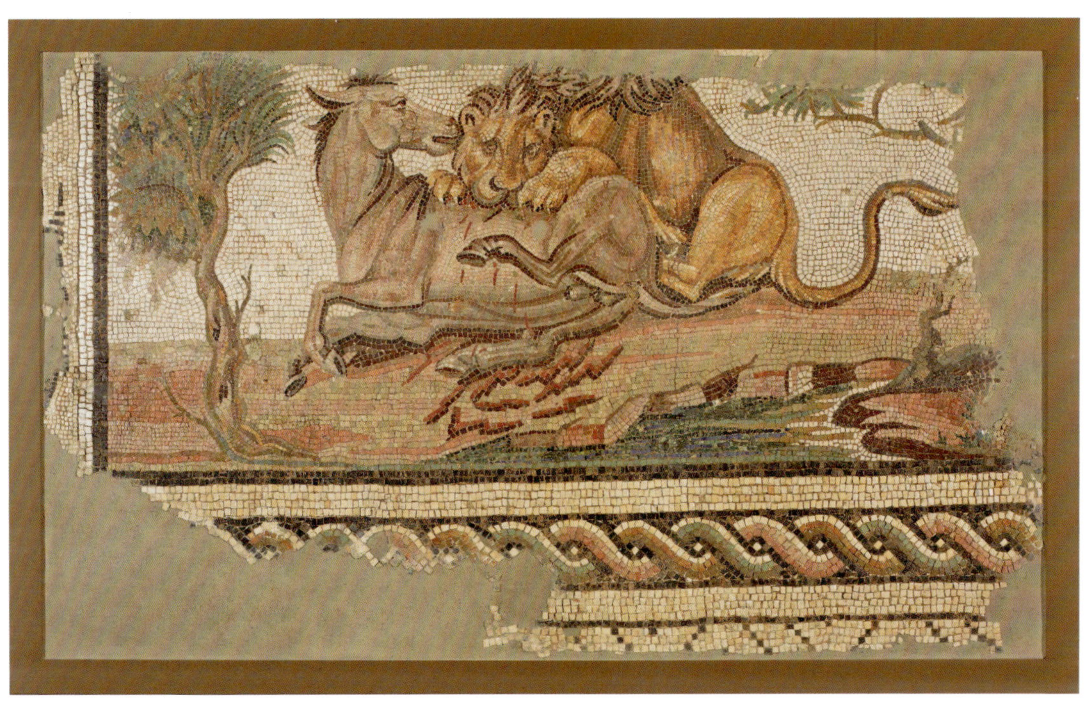

FIGURE 78
Cast mosaic snake. Roman, first century A.D. L 36 cm (14³⁄₁₆ in.). JPGM 2003.257.

FIGURE 79
Stone and glass mosaic depicting a lion killing an onager. Glass tesserae are used for some
of this scene's colors, such as the red blood of the onager. Roman, ca. A.D. 150. H 85.7 × L 146.8 cm
(33³⁄₄ × 57¹³⁄₁₆ in.). JPGM 73.AH.75.

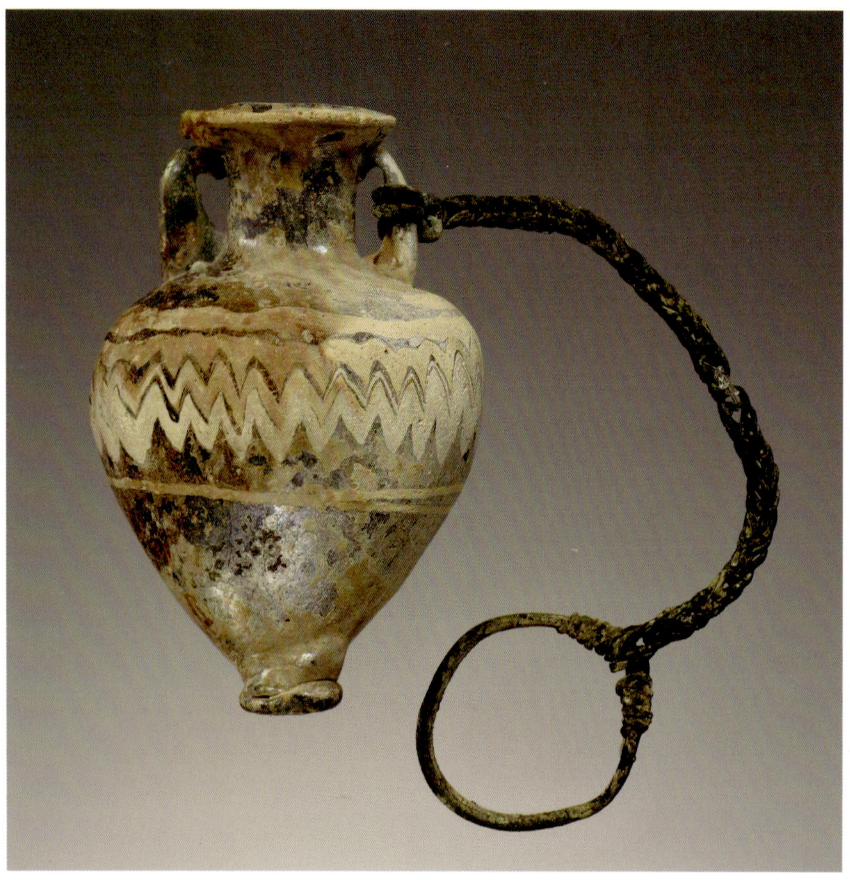

FIGURE 80
Core-formed perfume flask (amphoriskos) with a suspension chain. Greek, sixth–fourth century B.C. H 6 cm ($2\frac{3}{8}$ in.). JPGM 2003.168.

FIGURE 81
Ribbon perfume flask with tripod feet. Roman, first century A.D. H 11.3 cm ($4\frac{7}{16}$ in.). JPGM 2003.285.

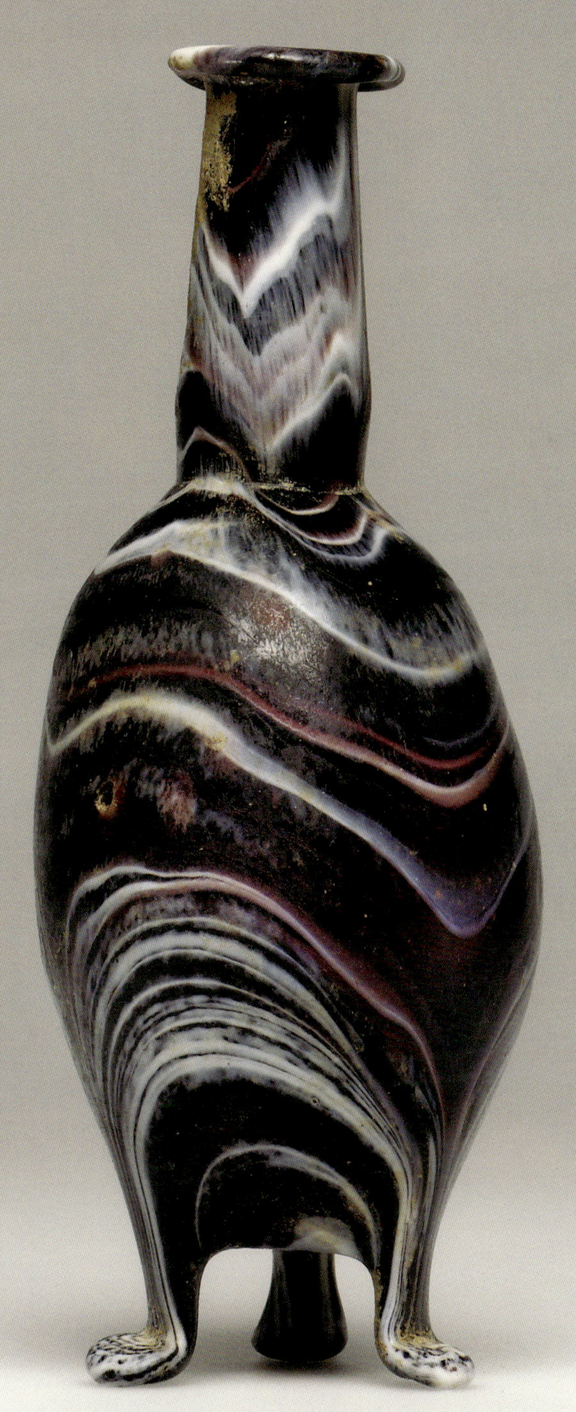

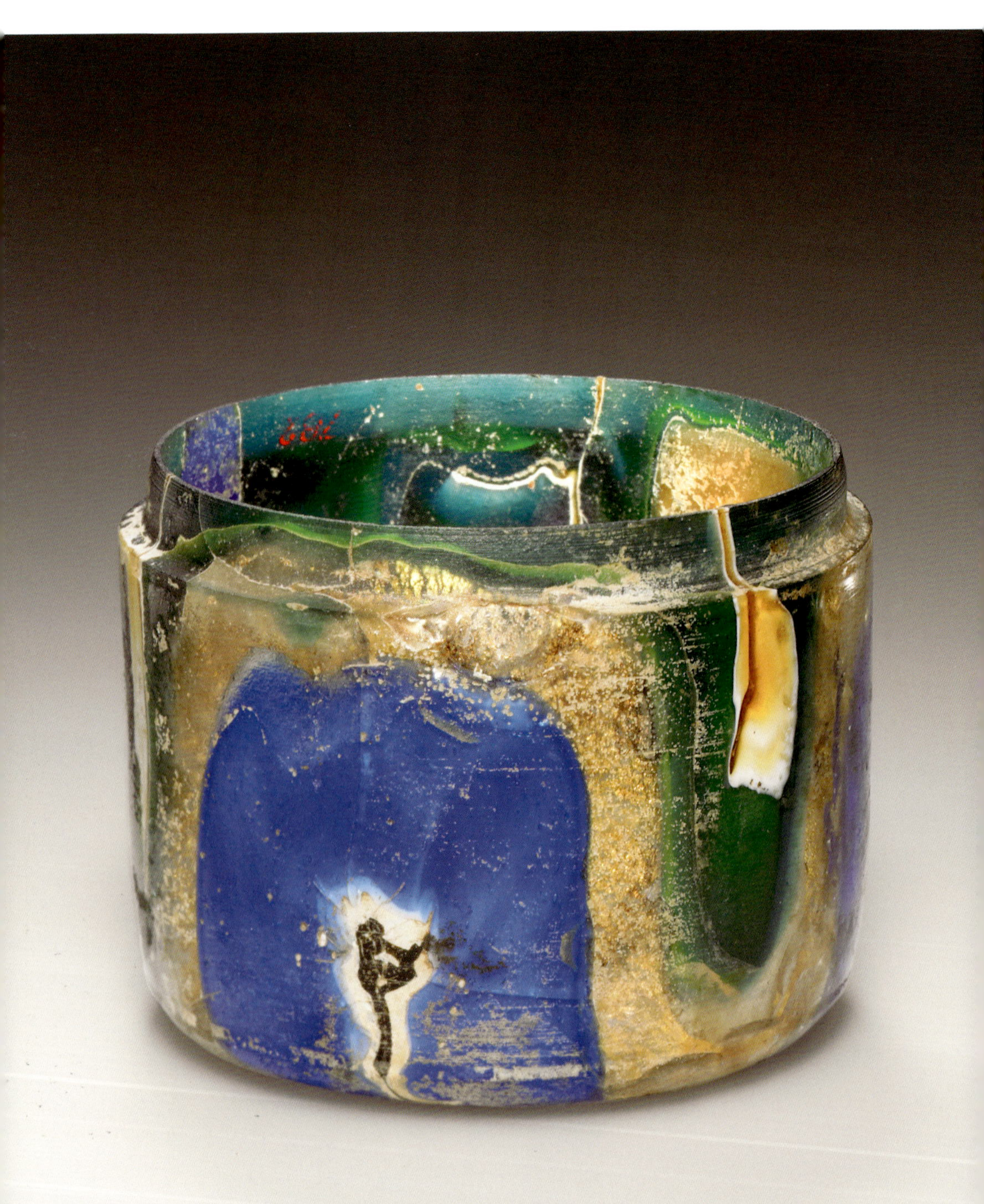

FIGURE 82
Gold band container (pyxis). Late Hellenistic or Roman, 25 B.C.–A.D. 25.
H 4.2 × DIAM 5.3 cm (1⁵⁄₈ × 2¹⁄₁₆ in.). JPGM 2003.231.

FIGURE 83
Kohl tube with bronze applicator rod. Roman, third–fourth century A.D. H 11.5 × W 7 cm
(4¹⁄₂ × 2³⁄₄ in.). JPGM 2003.431.

FIGURE 84
Rod for applying perfume. Rods such as this were used with the elongated alabastron shape
and were dipped inside the vessel in order to remove the contents for application to the skin.
Roman, first–second century A.D. H 18.5 cm (7⁵⁄₁₆ in.). JPGM 2003.403.

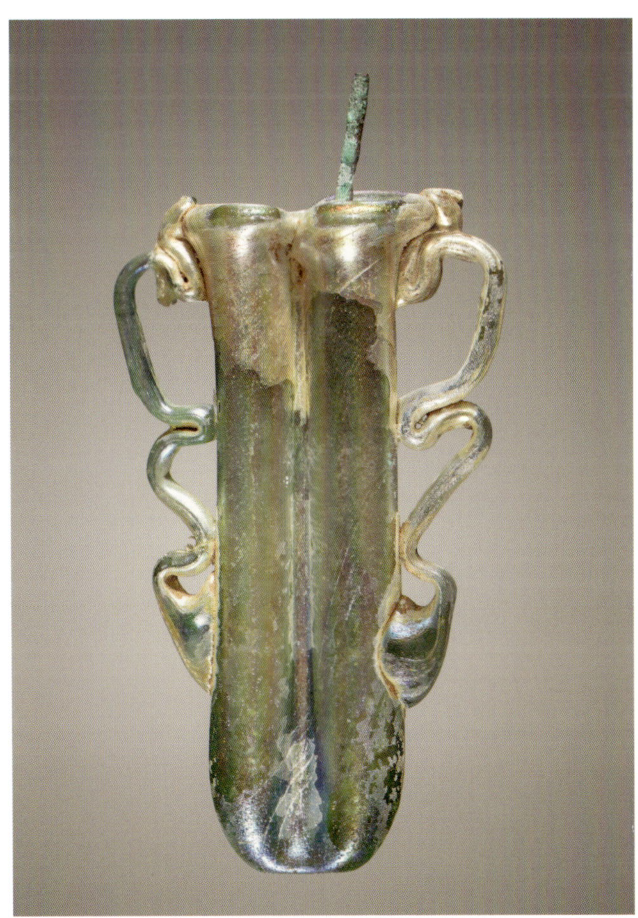

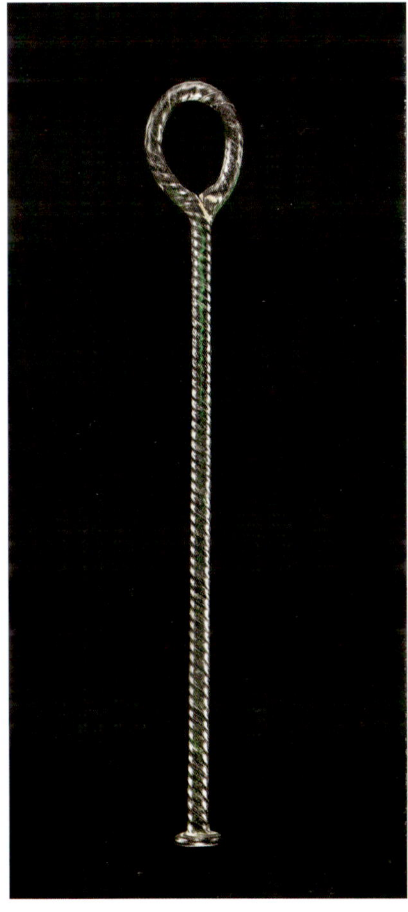

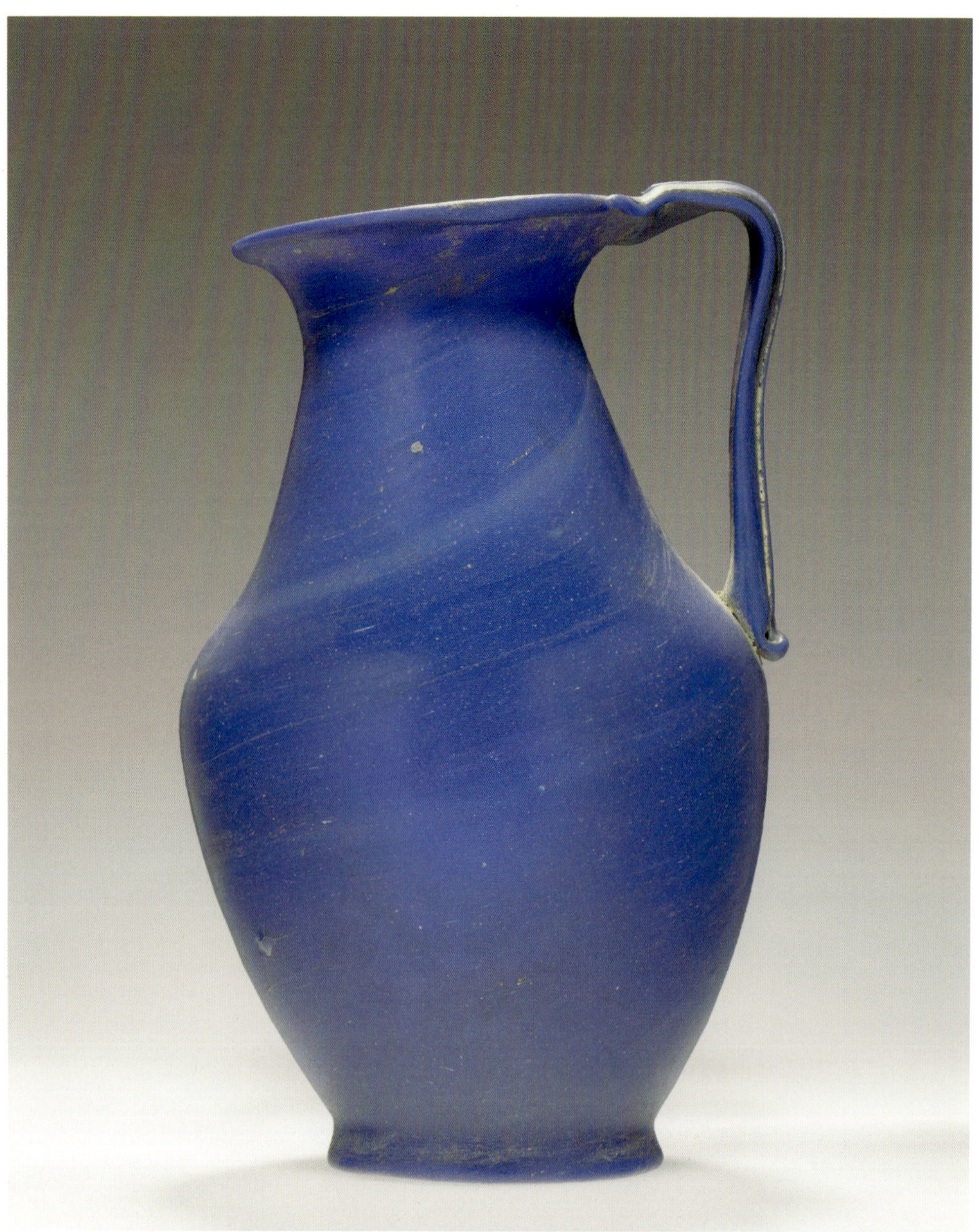

FIGURE 85
Pitcher (oinochoe). Roman, first century A.D. H 17 cm (6¹¹⁄₁₆ in.). JPGM 2003.476.

FIGURE 86
Two-piece beaker and drinking cup. Roman, first–second century A.D. H 8.2 × DIAM 8.2 cm
(3¼ × 3¼ in.). JPGM 2003.379.

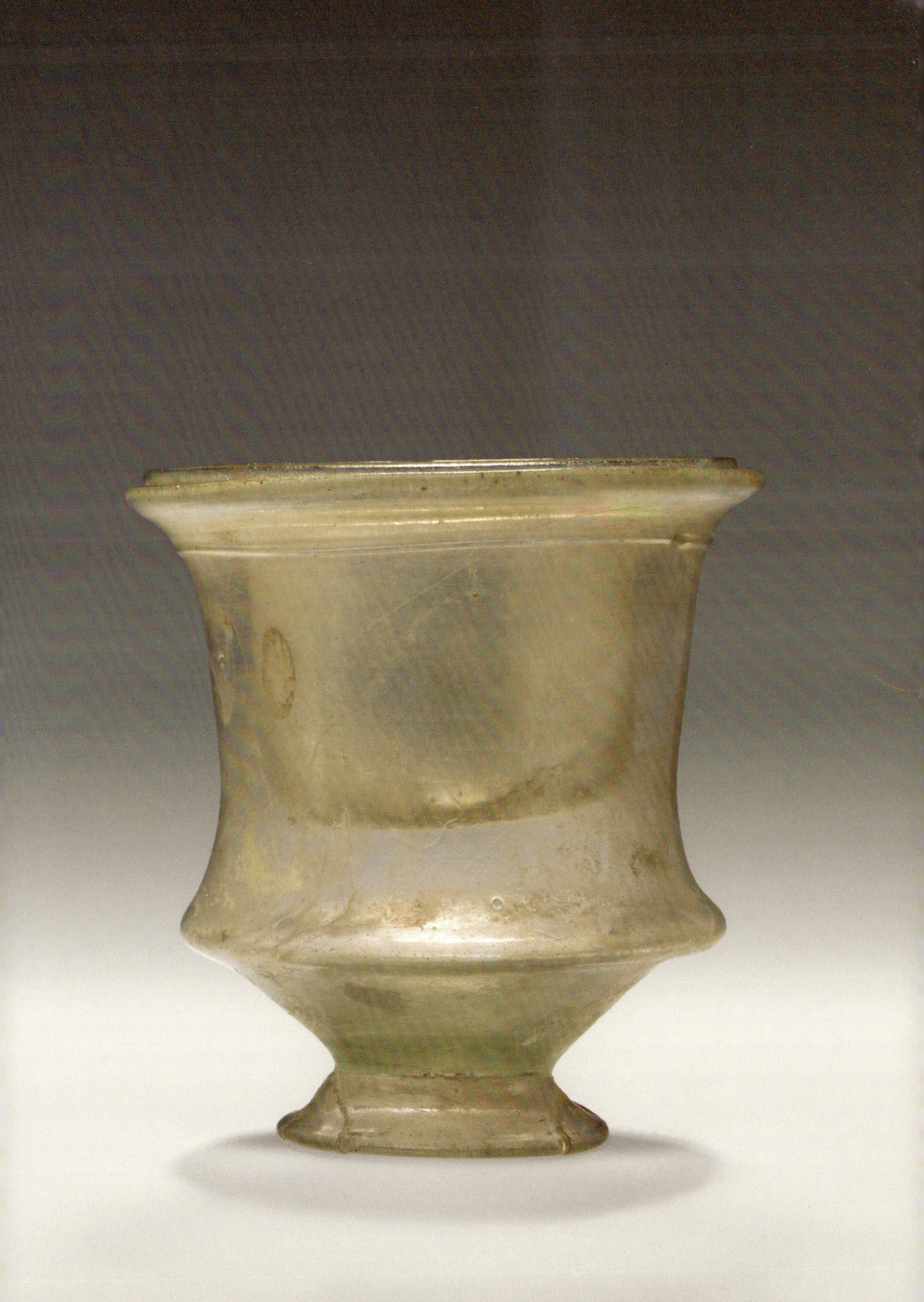

FIGURE 87
Plate. Roman, second–third century A.D. H 1.7 × DIAM 14.7 cm ($^{11}/_{16}$ × 5$^{13}/_{16}$ in.).
JPGM 2003.387.

FIGURE 88
Serving bowl with handle. Roman, third–fourth century A.D. H 16 × DIAM 11.5 cm
(6$^{5}/_{16}$ × 4$^{1}/_{2}$ in.). JPGM 2003.394.

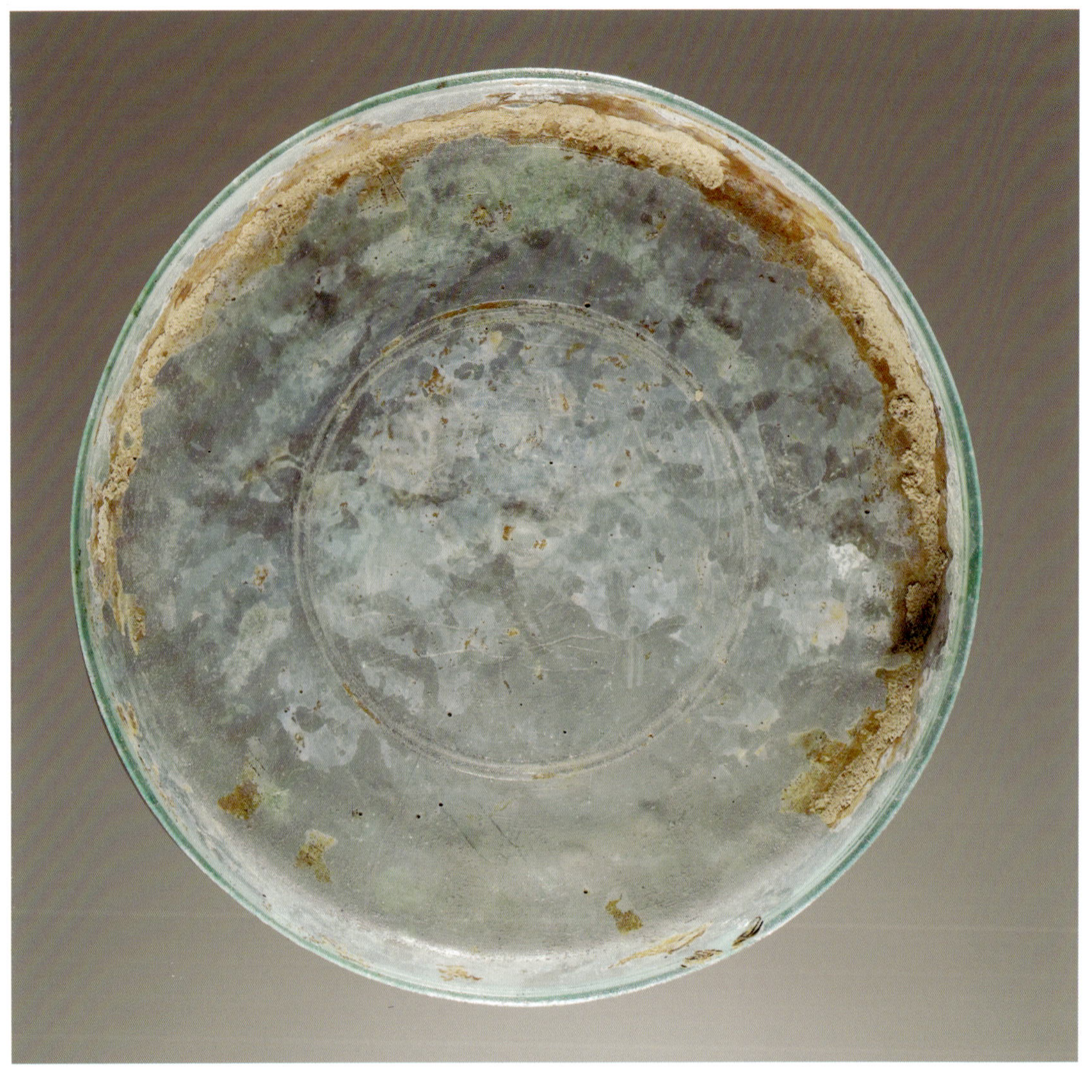

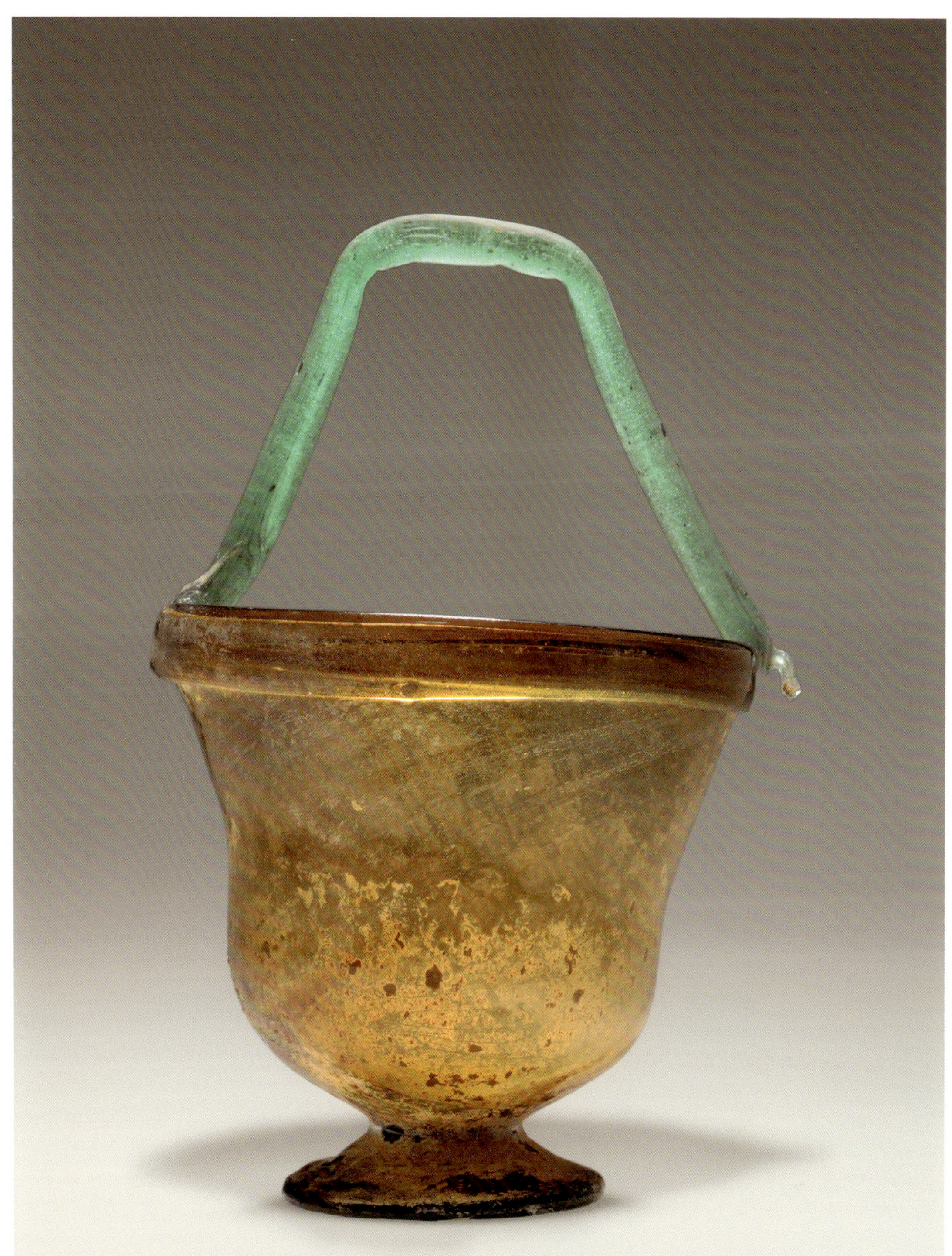

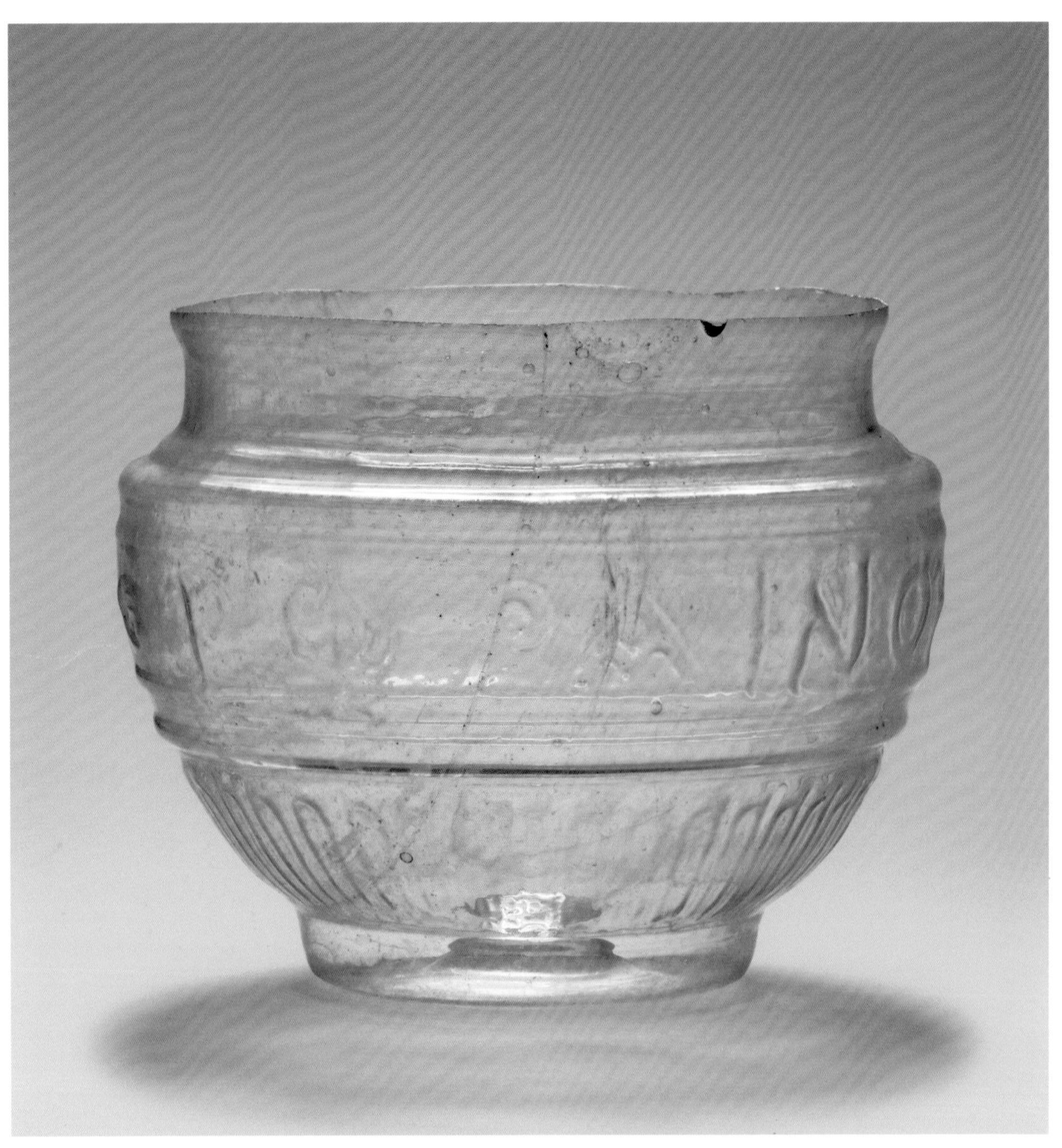

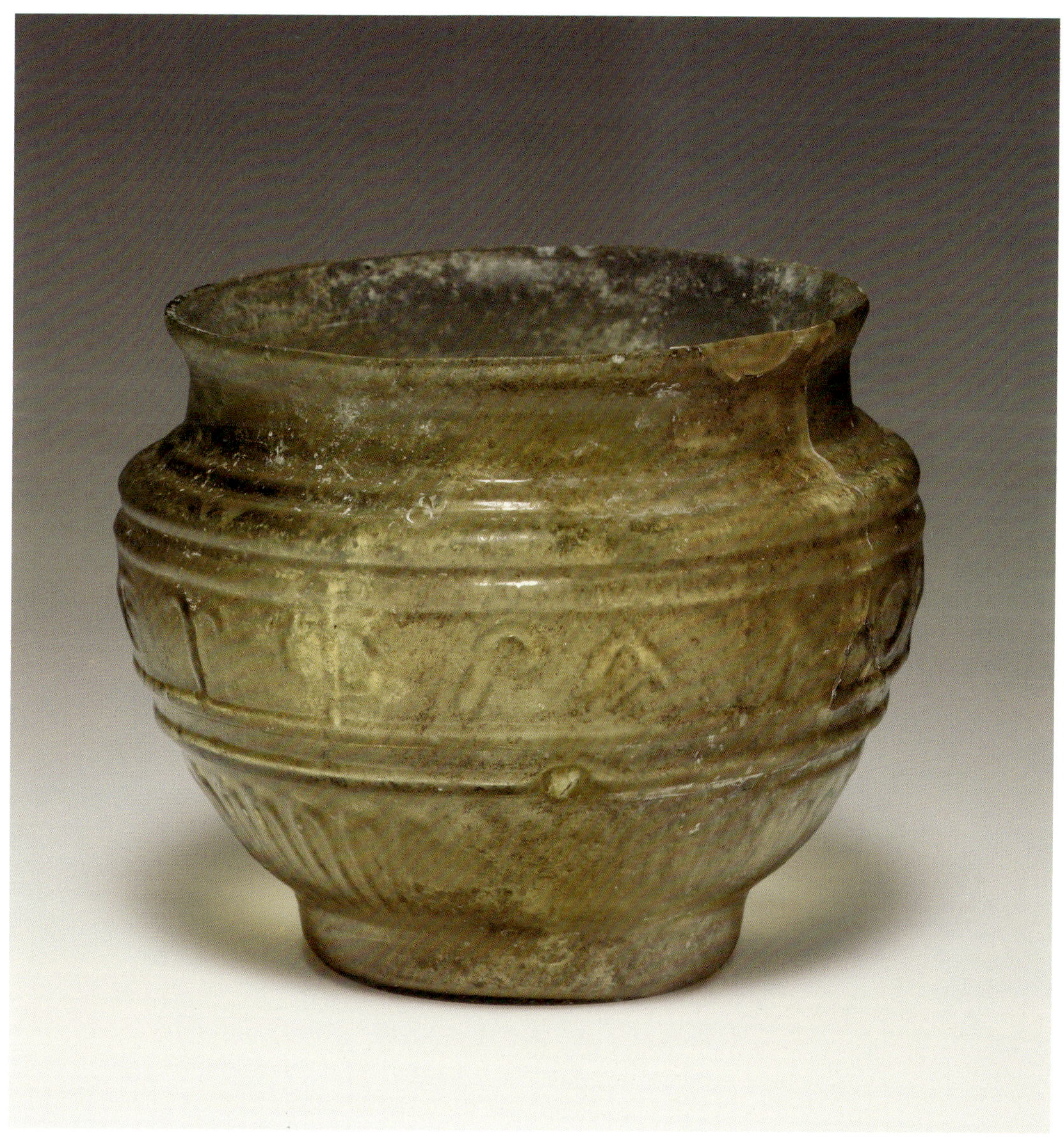

FIGURE 89

Mold-blown cup with an inscription. Roman, first half of the first century A.D. H 6.9 × DIAM (body) 8.7 cm (2¹¹⁄₁₆ × 3⁷⁄₁₆ in.). JPGM 95.AF.60.

FIGURE 90

Mold-blown cup with an inscription. Roman, first half of the first century A.D. H 7 × DIAM (at rim) 7.5 cm (2¾ × 2¹⁵⁄₁₆ in.). JPGM 2003.317.

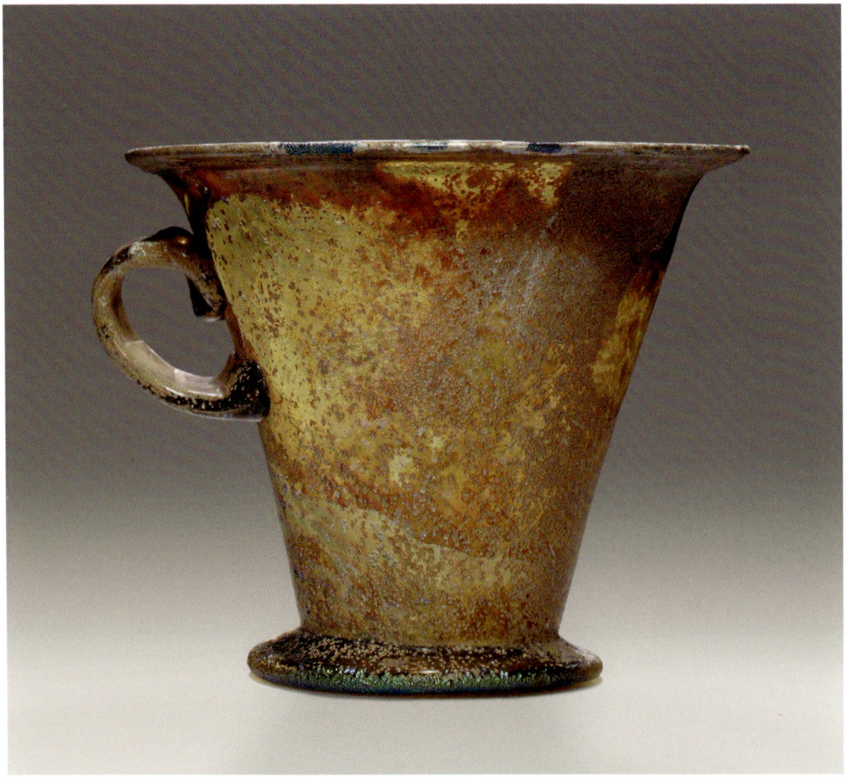

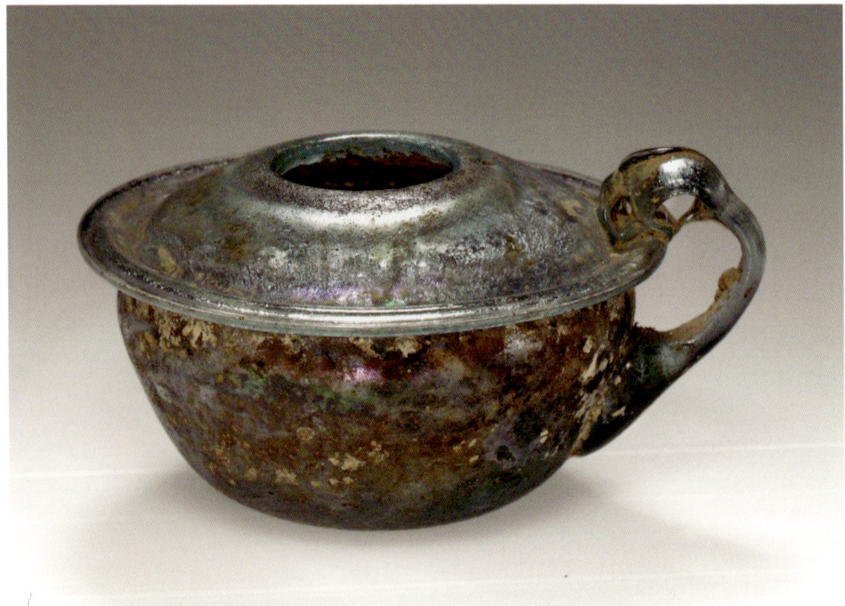

FIGURE 91
Measuring cup (modiolus). Roman, first–second century A.D. H 12.7 cm (5 in.).
JPGM 78.AF.31.

FIGURE 92
Oil lamp. Roman, first–second century A.D. H 3.3 × D 6.5 cm (1⁵⁄₁₆ × 2⁹⁄₁₆ in.).
JPGM 2003.378.

FIGURE 93
Cup from a hanging oil lamp (chandelier). Byzantine, fourth–fifth century A.D. H 7 × DIAM 10 cm
(2³⁄₄ × 3¹⁵⁄₁₆ in.). JPGM 2003.454.

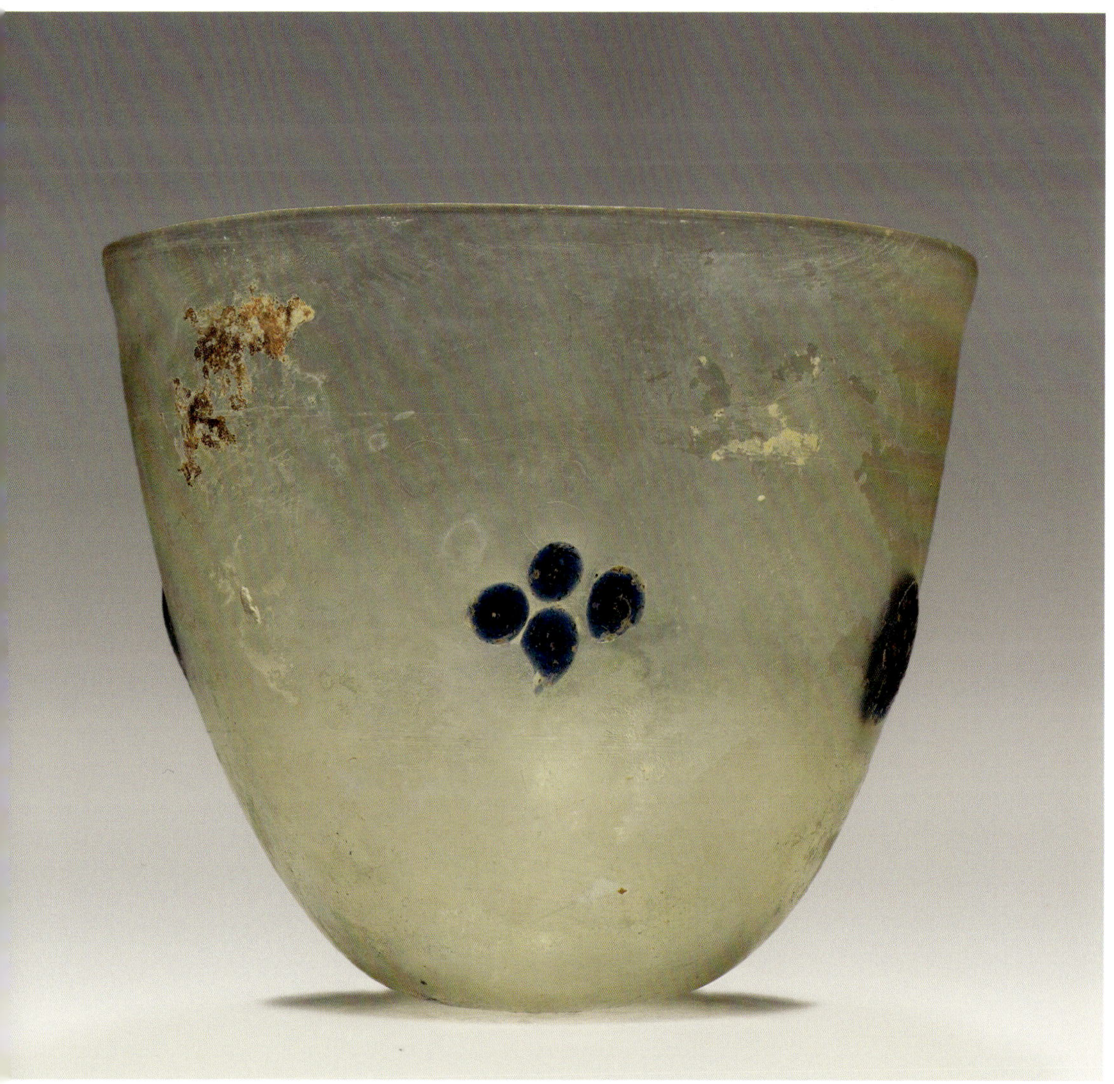

FIGURE 94
Spouted flask. Byzantine, sixth century A.D. H 15.8 × DIAM 7 cm (6¼ × 2¾ in.).
JPGM 2003.447.

FIGURE 95
Inkwell with lid. Late Hellenistic or Roman, first century B.C. With lid: H 9.5 × DIAM 6.3 cm
(3¾ × 2½ in.). JPGM 2003.239.

FIGURE 96
Cinerary urn. Roman, first–second century A.D. H 26 × W 25 cm (10¼ × 9¹³⁄₁₆ in.).
JPGM 80.AF.125.

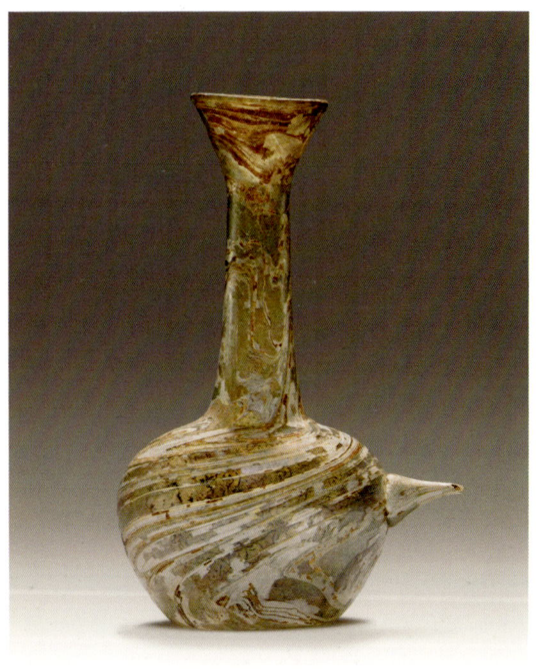

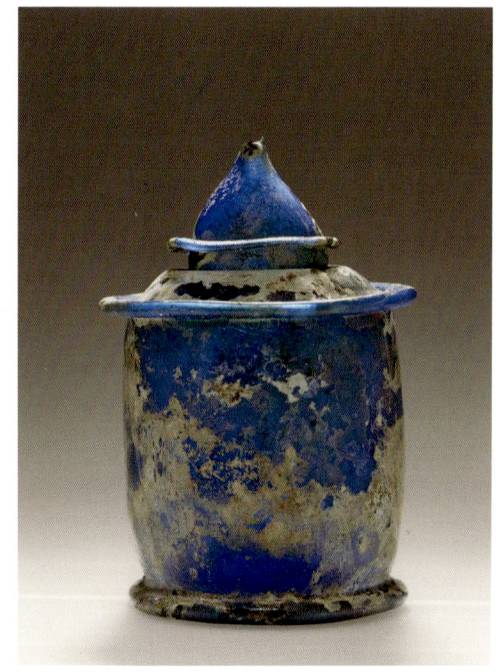

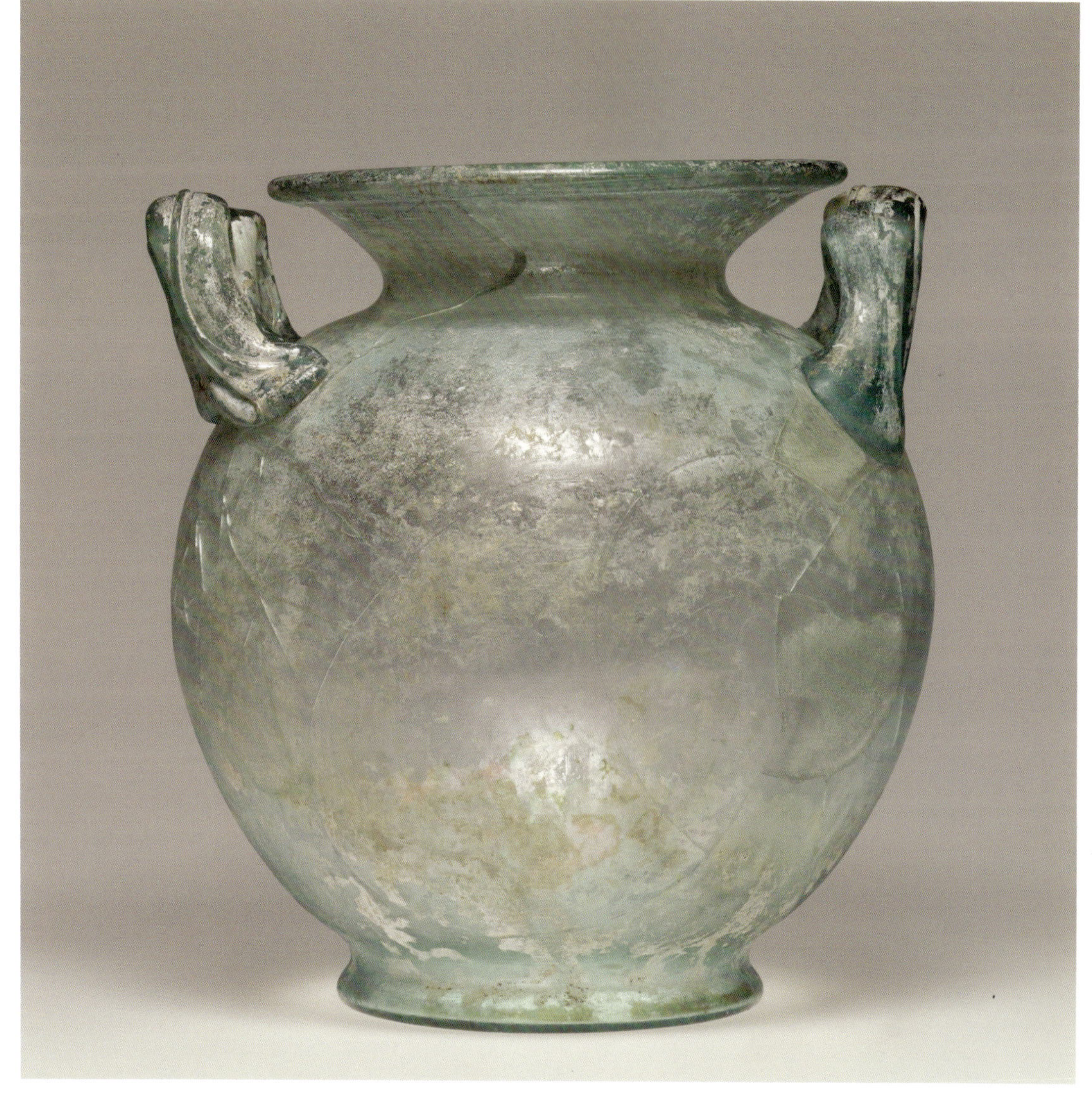

GLOSSARY OF GLASS TERMS

annealing
The process of heat treating and slowly cooling a completed object in a heated chamber known as a *lehr* (or *leer*) or *garage* in order to allow the stresses built up in the glass during manufacture to dissipate without deforming the object. This is a critical step in glassmaking, for if the hot glass object is allowed to cool too quickly, it will be highly strained by the time it reaches room temperature. Highly strained glass breaks easily if subjected to mechanical or thermal shock. The strain is caused by different parts of the glass cooling and hardening at different rates. To equalize the cooling process, the glass is initially heated to a uniform temperature and then allowed to cool uniformly in the lehr.

batch
The mixture of raw materials to make glass. The ingredients (see Figure 3) are heated in a crucible, or pot made of fireclay. Other elements, such as cullet and metallic oxide, can be added to the pot to form part of the batch. The mixture in a molten state is called the melt, although *batch* is commonly used to denote the same thing.

blowpipe
The iron or steel tube, usually about five feet long, used for inflating glass. Blowpipes have a mouthpiece on one end, and the other end flares out to form a conical shape on which the glass is gathered.

cane
A thin glass rod, usually monochrome, or a composite rod consisting of groups of single glass rods of different colors bundled together and fused to form a polychrome design that is visible when seen in cross-section. Canes with decorative finials were used in antiquity as perfume dippers and for stirring.

chair
The workbench used by the gaffer (glassmaker) while forming a glass object. Usually, this is a wide bench with arms, on which the gaffer rests the **blowpipe** or **pontil** with its parison (**gather**) of molten glass and rolls it back and forth so that the parison retains its symmetrical shape and stays on the blowpipe or pontil during the forming process. The term can also refer to the team of glassworkers who assist the gaffer.

dichroic glass
A type of glass whose physical properties enable it to appear one color in reflected light and another color when light is transmitted through it.

dip mold
A cylindrical, one-piece mold that is open at the top so that a **gather** can be dipped into it, inflated, and then removed.

enameling
Enamel is a substance composed of finely powdered glass that has been colored with a metallic oxide, mixed with an oily medium, and then applied to a glass object with a brush. The oily medium burns away during the firing, and the enamel fuses to the surface of the glass. Several firings are sometimes required to fuse several different enamel colors to a highly decorated object.

flux
An alkaline substance that lowers the melting point of materials in a mixture, thereby facilitating their fusion. For example, a flux is added to the **batch** in order to melt the silica, which normally requires very high and, therefore, hard-to-reach temperatures to fuse.

gather

A gob or mass of molten glass collected on the end of a tool such as a **blowpipe, pontil**, or gathering iron. The term also refers to the action of collecting the glass by rotating the iron while dipping its tip into the molten glass.

glory hole

Either an opening in the side of a glass furnace or a smaller, separate furnace used for reheating a hot glass object being produced or decorated. Inserting an object into the glory hole serves to fire-polish the piece, thereby melting and eliminating small surface imperfections.

jacks

A tonglike tool composed of two metal arms joined by a sprung section that is used to form the mouths of open vessels. The glassworker's grip controls the distance between the arms. Increasing the distance between the arms when the tool is placed inside the mouth of a vessel increases the aperture, while closing the tool on the outside of the piece makes the mouth narrower. Jacks are also used to shape stems and feet in particular and to adjust shapes of vessels in general.

marver

A smooth, flat surface onto which softened glass is rolled or pressed while attached to a **blowpipe** or **pontil**. This procedure is used to form and smooth the glass being worked and to pick up and consolidate applied decoration to the glass surface.

pontil

A solid metal rod to which a hot vessel is transferred from a **blowpipe** for further working. The vessel is attached to the vessel's base with a wad of hot glass.

shears

The heavy scissorlike tool used to trim away hot glass while it is attached to a **blowpipe** or **pontil**.

tempered glass

Also known as toughened glass or safety glass; glass that has been manufactured under controlled conditions in order to increase its strength. Using a thermal method, the glass is first heated to above 720°C, and then is chilled suddenly. This increases its tensile strength, making it hard to break, but if it should break, it crumbles into small pieces rather than shattering into larger fragments.

trail

A strand of glass drawn out from a **gather** and applied to another object; also called a thread.

These definitions have been borrowed from Catherine Hess and Karol Wight, *A Guide to Terms, Styles, and Techniques* (Los Angeles: J. Paul Getty Museum, 2005).

BIBLIOGRAPHY AND SUGGESTIONS FOR FURTHER READING

Stuart J. Fleming, *Roman Glass: Reflections on Cultural Change* (Philadelphia: University of Pennsylvania Museum of Archaeology and Anthropology, 1999).

Donald Harden, Hansgerd Hellenkemper, Kenneth Painter, and David Whitehouse, *Glass of the Caesars* (Milan: Olivetti, 1987).

Catherine Hess and Karol Wight, *Looking at Glass: A Guide to Terms, Styles, and Techniques* (Los Angeles: J. Paul Getty Museum, 2005).

Thilo Rehren and Edgar B. Pusch, "Late Bronze Age Glass Production at Qantir-Piramesses, Egypt," *Science* 308, no. 5729 (17 June 2005), 1756–58.

M. S. Walton, A. Shortland, S. Kirk, and P. Degryse, "Evidence for the Trade of Mesopotamian and Egyptian Glass to Mycenaean Greece," *Journal of Archaeological Science* 36 (2009), 1496–1503.

David Whitehouse et al., *Reflecting Antiquity: Modern Glass Inspired by Ancient Rome* (Corning, NY: Corning Museum of Glass, 2007).

CAST GLASS

Institute of Nautical Archaeology, Bodrum, Turkey. http://www.diveturkey.com/ inaturkey/ulub.htm/(accessed 20 October 2010).

Mark Taylor and David Hill, "Roman Glassmakers." http://www.romanglassmakers.co.uk/ (accessed 22 April 2010).

ANCIENT SOURCES IN THE LOEB CLASSICAL LIBRARY

Martial, *Epigrams,* trans. Walter C. A. Ker (Cambridge, MA: Harvard University Press, 1968).

Ovid, *Metamorphoses,* trans. Frank Justus Miller (Cambridge, MA: Harvard University Press, 1977).

Pliny the Elder, *Natural History,* trans. D. E. Eichholz (Cambridge, MA: Harvard University Press, 1967).

Seneca, *Investigations in Natural Science,* trans. Thomas H. Corcoran (Cambridge, MA: Harvard University Press, 1968).

Statius, *Silvae,* trans. D. R. Shackleton Bailey (Cambridge, MA: Harvard University Press, 2003).

Strabo, *The Geography of Strabo,* trans. Horace Leonard Jones (Cambridge, MA: Harvard University Press, 1966).

Suetonius, *Lives of the Caesars,* trans. J. C. Rolfe (Cambridge, MA: Harvard University Press, 1979).

INDEX